P9-DVJ-733

MAKING BABIES: BIOMEDICAL TECHNOLOGIES, REPRODUCTIVE ETHICS, AND PUBLIC POLICY

MAKING BABIES: BIOMEDICAL TECHNOLOGIES, REPRODUCTIVE ETHICS, AND PUBLIC POLICY

by

Inmaculada de Melo-Martín
St. Mary's University, San Antonio, Texas, U.S.A.

KLUWER ACADEMIC PUBLISHERS
DORDRECHT / BOSTON / LONDON

A C.I.P. Catalogue record for this book is available from the Library of Congress.

ISBN 0-7923-5116-9

Published by Kluwer Academic Publishers,
P.O. Box 17, 3300 AA Dordrecht, The Netherlands.

Sold and distributed in North, Central and South America
by Kluwer Academic Publishers,
101 Philip Drive, Norwell, MA 02061, U.S.A.

In all other countries, sold and distributed
by Kluwer Academic Publishers,
P.O. Box 322, 3300 AH Dordrecht, The Netherlands.

Printed on acid-free paper

Printed in the Netherlands

TO MY PARENTS ALFREDO AND CATALINA

MY BROTHER JAVIER AND MY SISTER BEATRIZ

WITH ALL MY LOVE

TABLE OF CONTENTS

CHAPTER 3

CHAPTER 4

CHAPTER 5

FOREWORD

Each year, roughly a million new cases of cancer appear in the US, and more than 500,000 Americans die annually of premature death. Although medical progress has slowed cancer mortality, its incidence is increasing roughly six times faster than cancer mortality is decreasing. Breast cancer, in particular, has been increasing about one percent each year since 1973. At least two of the factors responsible for this surge in breast cancer are women's use of medically-prescribed synthetic hormones and the exposure of the entire population to chemicals such as dioxin. Both exposures increase the likelihood of breast cancer. Although many ethicists worry about involuntary societal imposition of chemicals such as dioxin, through industrial and agricultural processes, allegedly voluntary exposures also constitute both, a public-health problem and a biomedical-ethics difficulty. Physicians recommend synthetic hormones, for example, to women who apparently take them voluntarily. In the case of *in vitro* fertilization, doctors prescribe hormones to induce egg production and to increase the chances of reproduction for couples who are unable to have children. Despite the benefits of medical technologies such as hormone stimulation and *in vitro* fertilization, they also carry great risks. The price that childless women pay, for their opportunity to have children through *in vitro* fertilization, may be their own increased risk of diseases – such as breast cancer – that are hormone dependent.

If women gave genuine informed consent to reproductive technologies that increased their cancer risks, then ethical problems with current use on *in vitro* fertilization would not be as significant as they are today. Nevertheless, it appears that fertility patients often have not given free informed consent to this set of techniques, in part because medical doctors have neither evaluated nor disclosed their attendant risks. In addiction, apart form the riskiness of the technologies and their partially involuntary imposition, *in vitro* fertilization and other reproductive technologies are neither equally available to all women nor use as sparingly as they ought to be. Indeed, they appear frequently to be overused and employed indiscriminately or without adequate justification. Such problems raise issues not only of biomedical and technological ethics but also of social and political philosophy.

Inmaculada de Melo-Martín's *Making Babies: Biomedical Technologies, Reproductive Ethics, and Public Policy* is the first book to deal extensively with the ethical, social, and epistemological problems raised with *in vitro* fertilization. It describes these problems in the context of the major assessments of the technology as used in major nations of the West, and it offers philosophical and policy solutions to their misuse and overuse. Unlike many philosophical accounts of technology, this volume avoids the anti-technological rhetoric of contemporary Luddites and the pro-technology bias of industry spokespersons. Instead, it offers a thoughtful assessment of a technology, like so many others, that has been used without adequate foresight and evaluation.

Making Babies is important not only because of the ethical and policy suggestions it offers to contemporary decisionmakers but also because of its applications of classical biomedical concepts to the problem of *in vitro* fertilization. As an analysis of a contemporary biomedical technology, the volume additionally is important because it is like a canary in a coal mine. Just as the death of a canary warned miners of methane before it reached levels lethal to humans, so also the threat posed by indiscriminate use of *in vitro* fertilization warns contemporary ethicists and policymakers. It warns them of the more general social problems of risk, discrimination, loss of equal opportunity, lack of free informed consent, and sexism that misuse of contemporary technologies brings. *Making Babies* functions not only as a superb canary, a wake-up call to society, but also as a good example of applied philosophy.

Kristin Shrader-Frechette
Tampa, Florida
January 1, 1998

ACKNOWLEDGEMENTS

This work would not have been possible without the help, inspiration, and loving support of Dr. Kristin Shrader-Frechette. I have not only found encouragement from her work, but Dr. Shrader-Frechette has provided me with a personal and professional role model. To her I owe my greatest debt.

I am grateful also to my friends and colleges Dan, Frank, Nancy, Peter, and Susan for their comments and suggestions at different philosophy seminars. Thank you also to Asun, Carlos, Miguel, Jason, and John for their help with this work and their encouragement through the writing process.

My parents, brother, and sister have been a continuous source of care, love, and support without which this work would have been impossible. Chus' trust and assurance have taken me to where I am now. Carl's patience and love have been invaluable during the completion of this manuscript.

I am also grateful to Jose Antonio Lopez Cerezo for his initial guidance through my studies at the University of Oviedo (Spain). My thanks also to the Spanish Government and to the Department of Philosophy of the University of South Florida for financial support for most of this work.

CHAPTER 1

INTRODUCTION: THE IMPORTANCE OF EVALUATING ASSESSMENTS OF IN VITRO FERTILIZATION

1.1. Introduction

Less than twenty years ago the first "test-tube baby" was born in England. Today, grandmothers serve as surrogate mothers for their daughter's children, postmenopausal women become pregnant, divorced couples fight over the custody of frozen embryos, and scientists clone sheep. The use and rapid proliferation of technologies such as *in vitro* fertilization (IVF) and related procedures have aroused public controversy around the world. Governments in different countries have established committees to analyze the social, legal, and ethical implications of IVF and associated techniques. A few legislatures have translated into law the recommendations put forward by their respective committees. However, most of these committees tend to see technology as lacking any inherent relationship with ethical issues.[1] When social and ethical problems arise, assessors understand them as secondary to the introduction of the technology and a result of the application of new abilities and instruments. In this decontextualized view, most of the problems with which evaluators deal are associated with the consequences of the implementation and use of new techniques. For example, the medical use of IVF brings about worries related to the disposition of embryos never reclaimed, the storage of spare embryos, issues associated with licensing of infertility clinics, and so on. IVF assessors also show concern with application criteria: should IVF be used for single mothers, unmarried couples, postmenopausal women? These questions are, of course, useful because they may help to demarcate the acceptable use of IVF in our society. However, focusing IVF

1. See, for example, H. ten Have, "Medical Technology Assessment and Ethics. Ambivalent Relations," *Hastings Center Report*, 25: 5 (1995): 13-19 (hereafter cited as Have, *Medical Technology*).

assessments exclusively on technical problems may make the analyses inadequate.

In limiting IVF evaluations to issues surrounding the technical details of IVF applications, assessors often have defined the social, ethical, and legal issues too narrowly. They typically have neglected an analysis of the social context in which the implementation of IVF is taking place, the balance between individual rights and the common good, or questions of allocation of scarce resources. Assessors also have disregarded the importance of a critical evaluation of the problem (infertility) IVF tries to solve or the impact of IVF and related technologies on women's lives.

1.2. Overview

Given the significance of the social, ethical, and legal aspects of the implementation and use of IVF for making adequate public policies, ignorance of them is likely to result in deficient political choices. The aim of this work is to argue that because of the inadequacies of IVF assessments, policy makers have made decisions about fertility treatments that may not be in the public's best interests. I shall focus on four IVF reports produced by institutional committees, respectively, in Australia, Spain, the United Kingdom, and the United States. In chapter two, an overview of technology assessment, its historical background, components, methodology, techniques, and philosophical presuppositions, introduces the relevance of the evaluation of new techniques. In this chapter, I offer a brief overview of technology assessment (TA) and analyze the concept of TA and some reasons for using it. I present a succinct historical background on the development of technology assessment practice and discuss ten components that TAs use in order to anticipate unforeseen higher-order impacts. In reviewing TA methodology, I focus in a brief description of analytic, synthetic, and empirical methods used in TA. I also discuss in chapter two one of the techniques that dominates the decisionmaking process in TA: risk-cost-benefit analysis (RCBA). Finally, I argue that TA and RCBA presuppose philosophical judgments, such as those about aggregation, partial quantification, and subjective preferences, that assessors should not neglect if they desire a critical and comprehensive evaluation of technology.

The third chapter offers a brief account of the development of IVF and related technologies. In order to contextualize the problems analyzed in this work, I start the chapter with a description of *in vitro* fertilization and embryo transfer procedure, usually abbreviated as IVF. I describe briefly the scientific development of IVF and give an account of some ethical dilemmas that clinical IVF raises in our society. Finally, I summarize some laws and regulations related to IVF and associated techniques in the United States, Australia, Canada, and several western European countries.

The next three chapters analyze some of the deficiencies of the Victorian, British, Spanish, and the United States IVF assessments. Attempting to provide constructive criticism of the inadequacies of these evaluations, I suggest some possible ways to overcome or ameliorate them. In chapter four, I show that evaluations of IVF are inadequate because, in ignoring epistemological and ethical problems such as choosing criteria for decisions under uncertainty, assessors have overlooked the possibility of jeopardizing women's health. I argue that evaluators have erred in their analyses because, underestimating both the existing scientific evidence and the insufficiency of data on IVF safety and effectiveness, they have condoned the use of IVF. Thus, they have implicitly sanctioned questionable criteria, such as expected-utility maximization, for deciding in a situation of scientific uncertainty. They also have preferred to minimize false positives over false negatives. As a consequence, many women may be exposed to needless risks. Likewise, I argue that, in sanctioning the expansion of IVF to an increasing number of reproductive conditions, assessors also have implicitly condoned questionable criteria for making decisions under uncertainty. In addition, I try to answer some of the possible criticisms to my arguments. One of those objections is that evaluators have recommended obtaining written informed consent from women as a way to overcome problems with risks, and therefore, people ought not to disagree if women freely choose to use hazardous and inefficient technologies. A second criticism refers to the fact that interfering with women's ability to choose risky technologies, such as IVF, might be paternalistic.

Chapter five argues that IVF assessments are problematic because they might have encouraged public policies that disregard the common good. I show that evaluators of IVF have erred in their analyses because, in presenting the problem of infertility as primarily an individual one, they have underestimated the role of social, ethical, and political solutions, such as prevention, in solving reproductive problems. Undervaluing these kinds of solutions is, however, problematic because it may underemphasize social influences on, and the community's responsibility for, the well being of its members and may foster unfair discrimination against women. Furthermore, I argue that IVF assessors have failed in their analyses because, in assuming too broad a definition of 'infertility', they may encourage people to use IVF needlessly. Thus, assessors seem to have conceded more importance to individuals' desires to have children than to problems of the common good. However, unnecessary use of IVF may threaten the common good because it might influence governments to increase needlessly the amount of resources dedicated to this technique. Promoting excessive funding of IVF is problematic because IVF and related technologies are expensive procedures. Given that health resources are scarce, allocating money for IVF could prevent the funding of other health-care measures that might be equally or more important. As in other chapters, I try to answer some possible objections to my arguments. First, critics may emphasize the difficulty of showing that social, ethical, and political solutions are cost-effective. Second, they may stress the fact that these kinds of answers to reproductive problems may require unattainable institutional changes. Third, they may argue that IVF evaluators have used the best available information on the prevalence of infertility. Fourth, critics may claim that IVF analysts have adopted the standard medical definition of infertility. Finally, critics may object that the definition of 'infertility' (used in the assessments) is more likely to enhance the welfare of greater number of individuals than other definitions are.

In chapter six, I deal with problems of free informed consent. I argue that assessors of IVF have failed because their evaluations might encourage public policies that jeopardize women's rights to free informed consent. First, I claim that because evaluators have underestimated problems with disclosure of information, their analyses may compromise women's opportunities to give free informed consent. Assessors have overlooked questions of disclosure because (i) they have underestimated the lack of scientific evidence on IVF safety, (ii) they have undervalued difficulties with the presentation of IVF success rates, and (iii) they have overemphasized the benefits of the procedure and have downplayed the hazards. Hence, analysts have skewed the balance of pros and cons and have made it difficult for women to give free informed consent to a risky treatment whose effects are uncertain. In addition, I argue that assessors have underestimated problems with voluntariness. Because evaluators have failed to analyze adequately social and economic conditions under which women make decisions about IVF, they have overlooked circumstances that could defeat women's rights to free informed consent. Finally, I respond to some potential objections to my arguments. For example, critics may say that assessors clearly advise doctors to obtain written informed consent forms from IVF patients. They also may argue that the task of IVF assessors is to analyze IVF and related procedures, not to offer value judgments about the context in which people implement and use these technologies.

1.3. The Four Assessments

The new assisted-conception technologies have raised many challenging ethical and policy issues in recent decades. Not only decisionmakers, but also medical practitioners, scientists, the courts, and the public in general are facing new quandaries that involve controversies among profoundly held values. When faced with ethical conflicts or complicated technical issues, policymakers often turn to or establish commissions for advice.[2] This has been the case with the new assisted-conception technologies.

---

2. See, for example, S. S. Connor and H. L. Fuenzalida-Puelma, eds., *Bioethics: Issues and Perspectives* (Washington, D.C.: Pan America Health Organization, 1990); Robert Blank, *Regulating Reproduction* (New York: Columbia University Press, 1990), chs. 5-7 (hereafter cited as Blanck, *Regulating)*; Office of Technology Assessment, *Biomedical Ethics in U.S. Public Policy* (Washington, D.C.: U.S. Government Printing Office, 1993); and Ruth Ellen Bulger *et al.*, eds., *Society's Choices. Social and Ethical Dimensions Making in Biomedicine* (Washington, D.C.: National Academy Press, 1995).

Commissions in several countries around the world have issued reports on IVF and related procedures.[3] The four analyzed in this chapter have, nevertheless, special significance for several reasons.[4] The Victorian report resulted in the first piece of legislation in the world to regulate assisted-conception techniques such as IVF. The British assessment has been the single most influential institutional inquiry on reproductive technologies and the one that set the agenda for action in other countries. The Spanish report has been the basis for Law No. 35/1988. This is one of the most detailed legislation undertaken on the subject of assisted-conception procedures. It covers artificial insemination, IVF, and gamete intrafallopian transfer (GIFT).[5] Finally, the United States assessment is the most complete of the four studies analyzed here. It covers the scientific, legal, economic, and ethical issues surrounding reproductive problems. Specifically, it evaluates medically assisted conception (including IVF and GIFT), surgically assisted reproduction, artificial insemination, basic research supporting reproductive technologies, and surrogate motherhood.[6]

Certainly, new reports and legislation have been passed in some of these countries.[7] However, the evaluation of these four reports is essential because they are, at least in part, responsible for the current situation in the development and use of IVF and other related technologies.

---

3. See, for example, Office of Technology Assessment, *Infertility: Medical and Social Choices* (Washington, D.C.: U.S. Government Printing Office, 1988)(hereafter cited as OTA, *Infertility)*; Blanck, *Regulating*; P. Singer *et al*., eds., *Embryo Experimentation* (Cambridge: Cambridge University Press, 1990) (hereafter cited as Singer, *Embryo*); and J. Gunning and V. English, *Human In Vitro Fertilization* (Aldershot: Dartmouth, 1993) (hereafter cited as: Gunning and English, *Human*). See, also, chapter Three for a description of some reports in different countries.

4. For a more complete account of these four reports see Chapter Three.

5. See Comisión Especial de Estudio de la Fecundación "In Vitro" y la Inseminación Artificial Humanas [Special Commission for the Study of Human in Vitro Fertilization and Artificial Insemination], *Informe* [*Report*] (Madrid: Gabinete de Publicaciones, 1987) (hereafter cited as Spanish Commission).

6. See OTA, *Infertility*, p.3.

7. See, for example, Singer, *Embryo;* and Gunning and English, *Human*.

The first of the assessments was published in the State of Victoria, Australia, in 1982. The Committee, under the direction of Louis Waller, produced three reports over two years, the first of which dealt specifically with IVF treatment. The Victorian Committee presented its *Interim Report* in September 1982, the *Report on Donor Gametes in Vitro Fertilization* in August 1983, and the *Report on the Disposition of Embryos Produced by In Vitro Fertilization* in August 1984. Following the recommendations of these reports, the Victorian Government enacted the *Infertility (Medical Procedures) Act* in 1984.[8] This bill sets out the provisions to regulate IVF and associated technologies. It allows the fertilization of ova outside a woman's body only for implantation and only for married couples. Counseling is mandatory. It limits the practice of IVF to approved hospitals and provides for record keeping and confidentiality.

In July 1982, two months after the appointment of the Victorian Committee, the British parliament established the Warnock Committee that produced the *Report of the Committee of Inquiry into Human Fertilization and Embryology*. Although both the Victorian and the British commissions reported their findings in 1984, the former led to immediate legislation, the latter to an extended period of consultation. Nevertheless, in 1990 the British parliament brought about the eventual recommendations of the Warnock proposals in the *Human Fertilization and Embryology Act*.[9] The Act establishes the statutory licensing of IVF, the donation and storage of eggs and sperm, and embryo research. It allows licensed research, but it does not permit the storage or use of embryos beyond 14 days of fertilization. The Act also prohibits cloning and the placing of a human embryo in any other animal.[10]

In 1986, the third assessment on new reproductive techniques under analysis appeared in Spain. The Spanish parliament set up a Special Commission to study human *in vitro* fertilization and artificial

8. See, for example, Singer, Embryo; and Gunning and English, *Human*.

9. See M. Warnock, *A Question of Life. The Warnock Report on Human Fertilization and Embryology* (Oxford, UK: Blackwell, 1985) (hereafter cited as: Warnock Report). See, also, Blanck, *Regulating*, p. 143; and Gunning and English, *Human*.

10. See Warnock Report. See, also, Blanck, *Regulating*, p. 143; and Gunning and English, *Human*.

insemination.[11] Following this report, the parliament passed *Health: Assisted Reproduction Techniques*. The law lays down general principles for the application of these technologies that emphasize informed consent, patient data collection and confidentiality, fertilization of ova for the sole purpose of procreation, and the minimization of spare embryos.

Finally, in 1988 the United States Office of Technology Assessment (OTA) issued its report *Infertility: Medical and Social Choices*. This assessment seems to have triggered the proposal of a federal law relating to IVF. After several hearings in 1988 and 1989,[12] the U.S. Congress enacted the *Fertility Clinic Success Rate and Certification Act of 1992*.[13] This Act requires all infertility clinics that perform IVF services to communicate annually their pregnancy success rates to the Secretary of Health and Human Services. It demands as well the identity of each embryo-laboratory working in association with the clinic. The Bill also directs the Secretary to develop a model program for State certification of embryo laboratory accreditation programs. In addition, it demands that the Secretary publishes and disseminates data concerning pregnancy success rates and other related information. The Centers for Disease Control and Prevention continue developing the actual mechanisms for the implementation of the Act.[14]

The main goal of these four committees was to provide direction for public policy in relation to IVF and other associated technologies. Unlike

---

11. See Spanish Commission.

12. See Subcommittee on Regulation and Business Opportunities, *Consumer Protection Issues Involving In Vitro Fertilization Clinics* (Washington, D.C.: U.S. Government Printing Office, 1988); Subcommittee on Regulation, Business Opportunities, and Energy *Consumer Protection Issues Involving In Vitro Fertilization Clinics* (Washington, D.C.: U.S. Government Printing Office, 1989).

13. See, Subcommittee on Health and the Environment, *Fertility Clinic Services* (Washington, D.C.: U.S. Government Printing Office, 1992); U.S. Congress, House of Representatives, *Fertility Clinic Success Rate and Certification Act of 1992* (Washington, D.C.: U.S. Government Printing Office, 1992), Report 102-624; and U.S. Senate, *Fertility Clinic Success Rate and Certification Act of 1992* (Washington, D.C.: U.S. Government Printing Office, 1992), Report 102-452.

14. See Centers for Disease Control and Prevention (CDC), *Assisted Reproducitve TechnologySuccess Rate in the United States:1995 National Summary and Fertility Clinic Report* (Atlanta, GA: Government Printing Office, 1997).

the Victorian, the British, and the Spanish studies, the United States report does not offer recommendations. Its main purpose is, however, similar to that of the other assessments: to help legislative policymakers in their tasks by providing information about the new reproductive technologies.

The main conclusions of these IVF assessments also are analogous, although there are some areas of disagreement. All of the reports have resolved that, at least in the case of married or stable couples, artificial insemination and *in vitro* fertilization are legitimate medical responses to the problem of infertility. They agree that informed consent is a precondition for treatment with these procedures. They argue that some forms of embryo research, such as cloning, clearly are unacceptable; others are permissible within the first 14 days of development *in vitro*, provided that ethics committees regulate and approve them. Commissioners also agree that governments should allow the donation of embryos. Similarly, the reports concur that governments should regularize the legal status of children conceived through the new reproductive technologies. They also emphasize the need to establish some form of national accreditation or licensing for assisted-conception clinics.

1.4. Previous Work on in Vitro Fertilization and Related Technologies

Since the first attempts to fertilize human eggs *in vitro*, the new assisted-conception techniques have generated both significant debate and important literature.[15] The available bibliography on reproductive procedures may be divided into three major groups: technical works, discussions of the ethics of embryo experimentation, and feminist analyses.[16] The first group includes the works of the "experts" on the new procedures. These are technical descriptions that laud the remarkable power of modern science and technology and celebrate the possibility of

15. See chapter three for a historical overview of social, legal and ethical issues arising from the use of IVF and related technologies.

16. See, for example, L. Birke, S. Himmelweith, and G. Vines, *Tomorrow's Child* (London: Virago Press, 1990) (hereafter cited as Birke, Himmelweith, and Vines, *Child*).

solving the problems of people with reproductive difficulties.[17] In these works, IVF and related techniques are not questioned but presented as solutions to infertility problems. Often, these studies do not inform women about the health risks associated with these procedures.

The second group of studies on IVF focuses on ethical issues arising from embryo experimentation. Ethical and religious questions about personhood, the rights of embryos, cloning of humans, and ownership of embryos are the main topics of these works. In general, they disregard both the ways in which IVF and related techniques affect women and the ethical issues associated with clinical use of IVF.[18]

The third group of studies on assisted conception techniques stresses women's rights and needs. Feminist analyses in this category adopt a more women-centered position and express, in some cases, opposition to these techniques.[19] In others instances, they recommend caution.[20] These works

---

17. See, for example, R. Edwards and P. Steptoe, *A Matter of Life: The Story of a Medical Breakthrough* (London: Hutchinson & Co., 1980); and A. Trouson and C. Wood, eds., *In Vitro Fertilization and Embryo Transfer* (Edinburgh: Churchill Livingstone, 1984).

18. See, for example, P. Ramsey, *Fabricated Man: The Ethics of Genetic Control* (New Haven: Yale University Press, 1970); L. R. Kass, " Babies by Means of In Vitro Fertilization: Unethical Experiments on the Unborn? *New England Journal of Medicine* 285 (1971): 1174-1179; J. D. Watson, "Potential Consequences of Experimentation with Human Eggs," in Committee on Science and Astronautics, *International Science Policy* (Washington, D.C.: U.S. Government Printing Office, 1971); P. Ramsey, "Shall We `Reproduce'? I. The Medical Ethics of In Vitro Fertilization," *JAMA* 220:10 (1972): 1346-1350; P. Ramsey, "Shall We `Reproduce'? II. Rejoinders and Future Forecast," *JAMA* 220:11 (1972): 1480-1485; L. R. Kass, "Making Babies --the New biology and the `Old' Morality," *The Public Interest* 26 (1972): 18-56; M. Lappé, "Risk-Taking for the Unborn," *Hastings Center Report* 2:1 (1972); and Congregation for the Doctrine of the Faith, *Instruction of Respect for Human Life in its Origin and on the Dignity of Procreation* (Vatican City: Vatican Polyglot Press, 1987).

19. See, for example, R. Arditti, R. Klein, and S. Minden, eds., *Test-Tube Women. What Future for Motherhood?* (London: Pandora Press, 1984); G. Corea, *The Mother Machine: Reproductive Technologies from Artificial Insemination to Artificial Wombs* (New York: Harper and Row, 1985); G. Corea *et al.*, eds., *Man-Made Women. How Reproductive Technologies Affect Women* (London: Hutchinson, 1985); J. A. Scutt, ed., *Baby Machine. Reproductive technology and the Commercialization of Motherhood* (Melbourne: McCulloch Publishing, 1988); P. Spallone, *Beyond Conception: The New Politics of Reproduction* (London: Macmillan, 1989); R. Klein, *The Exploitation of a*

attend to the social context in which the new reproductive technologies exist, and to the ways these techniques may affect women as individuals and as a group.

Works in these three groups have served to call attention to the technical, ethical, social, and legal issues associated with IVF and related procedures. Nevertheless, few of them have analyzed the relevance of institutional assessments of these techniques for the implementation of particular policies that may not be in the best interest of the public. To the best of my knowledge, there is no philosophical work evaluating the assessments I analyze here. Some authors have dealt with particular aspects of some of these reports such as debates over embryo experimentation.[21] I do not know, however, of any investigation that debates the problematic evaluation of IVF in the Victorian, British, Spanish, and the United States assessments and its implications for public policy in relation to infertility treatments. Thus, this work is new for several reasons. It examines epistemological and ethical problems in important institutional assessments of IVF. It also evaluates the possible consequences for public policy on infertility treatments of inadequate technology analyses.

1.5. Analyzing in Vitro Fertilization Assessments: Importance

The growing concern of the public with the impacts of new technologies in our society has given place to an increase in the number of technology

---

Desire. Women's Experiences with In Vitro Fertilization (Geelong: Deaking University Press, 1989); R. Rowland, *Living Laboratories. Women and the New Reproductive Technologies* (Bloomington: Indiana University Press, 1992), R. Klein, *The Ultimate Colonization: Reproductive and Genetic Engineering* (Dublin: Attic Press, 1992).

20. See, for example, M. Stanworth, ed., *Reproductive Technologies. Gender, Motherhood, and Medicine* (Minneapolis: University of Minnesota Press, 1987); C. Overall, *Ethics and Human Reproduction. A Feminist Analysis* (Boston: Allen & Unwin, 1987); C. Overall, ed., *The Future of Human Reproduction* (Toronto: The Women's Press, 1989); and Birke, Himmelweith, and Vines, *Child.*

21. See B. Gaze and P. Kasimba, "Embryo Experimentation: The Path and Problems of Legislation in Victoria," in Singer, *Embryo*

assessments (TA).[22] Nevertheless, in spite of interest in the social, ethical, and legal implications of technological development, few TAs include a systematic analysis of these implications as a central and substantial element.[23] Such failures are particularly problematic because, in many cases, governments use these studies to guide formulation of public policies. Hence, if the assessments are inadequate, it is likely that public policies will reflect the deficiencies.

1.5.1. RELEVANCE FOR PHILOSOPHY

The present work tries to show that problematic evaluations of new techniques may be as dangerous as no evaluations at all. Because technology assessments may give people a false sense of security (experts have evaluated the social, political, ethical, and technical consequences of a new procedure), poor ones may serve as a legitimization of premature, dangerous, or misunderstood decisions. Careful analyses of technology assessments may uncover particular epistemological or ethical problems that may misguide public policy.

Because of their training, philosophers have an important role not only as analysts of problematic assessments, but also as contributors to technology evaluations.[24] Philosophers are trained to understand logical reasoning, detect fallacies, uncover hidden assumptions, show unexpected consequences of particular proposals, identify and articulate precisely the value dimensions of specific situations, analyze concepts and clarify meanings, and recognize normative, epistemological, and social issues. Of course, knowledge of values, meanings, and valid argumentation is not sufficient to guarantee the resolution of ethical problems related to a particular technology. However, knowledge of moral principles and theories, understanding of ethical and epistemological concepts, ability to argue clearly and rigorously, and knowledge of relevant facts are important

---

22. See Chapter Two for references.

23. See Have, Medical Technology.

24. See, for example, K. Shrader-Frechette, *Science Policy, Ethics, and Economic Methodology* (Dordrecht: Reidel, 1985), ch.1; and R. W. Momeyer, "Philosophers and the Public Policy Process: Inside, Outside, or Nowhere at All?, *The Journal of Medicine and Philosophy*, 15: 4 (1990): 391-409.

features for any analysis. Thus, insofar as these characteristics are significant in evaluating new procedures and in designing public policies, to that extent, philosophers are well prepared to participate in, and analyze, technology assessments.

The present study is also important for philosophy because it focuses on topics often neglected in analyses of IVF. For example, evaluations of this and related procedures generally have ignored the importance of IVF assessments in the selection of public policies. As I said in the previous section, there are no philosophical works evaluating the assessments I discuss here. Because governments have used institutional IVF studies to propose public policies in relation to infertility treatments, evaluation of IVF analyses seems especially relevant.

The present work is significant, as well, because it centers around an often neglected topic in philosophy: how IVF and related procedures affect women. Although women are the only ones who undergo IVF and other infertility treatments, concern about them is usually absent from evaluations of infertility procedures. Because women are the ones who bear and give birth to children, the implementation and use of any reproductive technology would surely affect them. Therefore, an analysis of how IVF and related procedures may influence women seems necessary. Focusing on women means not only evaluating how these technologies may harm or benefit particular individuals, but also how they may affect the role of women in society and women as a group. It also means taking into account social, political, and economic situations that may favor the application and use of IVF, may increase the difficulties of being childless, may prevent women from giving free informed consent, or may favor discrimination and unethical experimentation on women.

1.5.2. SIGNIFICANCE FOR PUBLIC POLICY

Public policy related to IVF does not affect only those who are infertile; it affects all of us. The introduction and use of these technologies have the potential to affect our conceptions of the family, our ideas about personhood, and the role of women in society. The use of IVF is increasing; the pool of candidates for whom doctors indicate this and related procedures is expanding; and the ethical and political dilemmas (i.e., access to IVF, allocation of scarce resources, informed consent problems, burdens and benefits) require our attention. An analysis of epistemological and ethical problems in IVF assessments may prevent decisionmakers from choosing policies that may not be in the public's best interests. It also may help to preclude evaluators from committing the same mistakes when examining other technologies. Better assessments hold considerable promise for better decisions. Our values shape the kinds of technologies that we implement and use. But new procedures also mold the values we support. There is an inextricable relation between technology and social context. When evaluating new techniques, assessors should take into account this relationship and question how the technologies we develop affect the kind of society we should have.

The use of new assisted-conception techniques has the potential to transform many of our values about personhood, the family, or women. A careful analysis of these procedures, the context in which they are implemented, and the reasons for their utilization may help decisionmakers and the public to prevent unethical use of IVF and related technologies. Failing to do such an analysis, may produce disastrous consequences for women in particular and for our communities in general.

CHAPTER 2

OVERVIEW OF TECHNOLOGY ASSESSMENT

1.1. Introduction

When television came into our living room, few people could predict that it would become a popular nanny. Similarly, not many people foresaw that the automobile would become a major contributor to air pollution and one of the principal causes of premature death.[1] Technology has grown into a major force shaping our society.

Although the effects of a new technology may be so devastating that its evaluation seems necessary, some conceptions of technology have hindered analysis of its impacts. This is the case with the image of technology as value neutral and autonomous. Those who view technology as value neutral see it as a collection of machines, techniques, and tools.[2] They perceive it as a result, not as a process: the result of applying "pure" science. From this point of view, technology as such does not incorporate any political or social values: It is merely a passive tool. According to this position, users can employ artifacts for good or evil, but the artifact is neither good nor evil in itself. On this view, if users employ technology in an improper way, the fault lies with them, not with it. From this position, users of technology, not scientists and engineers, are responsible for its damaging applications. Supporters of this thesis recognize that technology

---

1. Carbon dioxide emissions from motor vehicles account for about one fourth of total U.S. carbon dioxide emissions. See Committee on Science, Space, and Technology, *The Environmental Costs of Transportation Energy use* (Washington, D.C.: U.S. Government Printing Office, 1992), p. 37. Also, motor vehicle crashes are the leading cause of death for every age from 6 through 33 years old. See National Highway Traffic Safety Administration's National Center for Statistics and Analysis, *1992 Traffic Safety Facts* (Washington, D.C.: U.S. Department of Transportation, 1993).

2. See, for example, M. Bunge, "Five Buds of Techno-Philosophy," *Technology in Society* 1 (1979).

certainly can have harmful effects, such as pollution. However, they argue that those impacts arise from faulty applications, wrong social policy or a lack of sophistication, rather than from the technology itself.

The conception of technologies as value-neutral, however, is problematic.[3] First, the value-neutral thesis overlooks the social, economic, and policy assumptions of those who design, fund, build, and implement technologies. It ignores the fact that they can exhibit political qualities.[4] In some situations their design, invention, or arrangement can support particular social values. For example, an artifact as apparently innocuous as a bridge can be politically burdened. This is the case of the bridges over the parkways on Long Island, New York. Because Robert Moses, the builder, intentionally designed and constructed many of the overpasses extremely low, he prevented bus traffic on Long Island. His purpose was to limit access of poor people and blacks, who normally used public transport, to the parks and beaches of the area.[5] In other cases, technologies can be, by their very nature, political in the sense that they are more compatible with certain models of authority and power. To employ them is to choose a particular form of political life. For example, to base the energy supply of a country on nuclear power is also to create a centralized, rigid, and hierarchical social structure. Because of the highly dangerous weaponry possibilities of its byproducts such as plutonium, and the toxicity of uranium, nuclear power must be carefully controlled. Solar energy, however, is more compatible with a decentralized and non-hierarchical social system.[6]

---

3. For a critique of the thesis of the neutrality of technology see, for example, D. Dickson, *Alternative Technology and the Politics of Technical Change* (New York: University Books, 1974); L. Winner, *The Whale and the Reactor*, (Chicago: The University of Chicago Press, 1986)(hereafter cited as Winner, *The Whale)*; H. Oberdiek, "Technology: Autonomous or Neutral", *International Studies in the Philosophy of Science* 4, (1990): 67-77 (hereafter cited as Oberdiek, *Technology*); A. Feenberg, *Critical Theory of Technology* (New York: Oxford University Press, 1991); and R. D. Bruner and W. Ascher, "Science and Social Responsibility," *Policy Sciences* 25 (1992): 295-331.

4. Winner, *The Whale*, ch. 2.

5. Winner, *The Whale*, pp. 22-25.

6. Winner, *The Whale*, pp. 29-38.

A second problem with the conception of technology as value-neutral is that it discourages democratic participation in scientific and technological policy making. If technology is value free, then the only important factors in its evaluation are its scientific and technical characteristics. As a consequence, only the opinions of scientists and engineers, not the ideas of the layperson, are essential to technology assessment.[7] Thus the value-neutrality thesis obstructs critical analysis by the public and opens the door to a technocratic policy, i.e., decisionmaking by scientific and technical elite.[8] If society realizes that technology is not neutral but value laden, then society in general, and not only the "experts," should be able to accept or reject it. Policy about technology concerns everyone. Experts may and should have a voice in its assessment; they do not, however, have the right to impose their evaluative preferences about technology under the assumption that they are making only a technical decision--at least not without taking into account societal values.

Another conception that has helped constrict improperly the evaluation of technology is the thesis of its autonomy.[9] The autonomy belief results

---

7. For a criticism of the actual role of the experts in the technological decision making process see, for example, S. Jasanoff, *The Fifth Branch: Science Advisors as Policymakers* (Cambridge, MA: Harvard University Press, 1990) (hereafter cited as: Jasanoff, *The Fifth Branch)*; K. Shrader-Frechette, *Risk and Rationality* (Berkeley, CA: University of California Press, 1991)(hereafter cited as: Shrader-Frechette, *Risk)*; B. Wynne, "Misunderstood Misunderstanding: Social Identities and the Public Uptake of Science," *Public Understanding of Science* 1(1992): 281-304; K. Shrader-Frechette, "Probabilistic Uncertainty and Technological Risks," in *Philosophy and Methodology of the Social Sciences*, ed. R. von Schomber, vol. 17, (Boston: Kluwer, 1993), pp. 43-62.

8. See J. Habermas, *Toward a Rational Society* (Boston: Beacon Press, 1970); Jasanoff, *The Fifth Branch*; N. Postman, *Technopoly*, (New York: Alfred A. Knopf, 1992).

9. For a defense of the autonomy of technology thesis see, for example, J. Ellul, *The Technological Society* (New York: Alfred A. Knopf, 1964)(hereafter cited as: Ellul, *Technological Society*); H. Marcuse, *One Dimensional Man* (Boston: Beacon Press, 1964); H. Jonas, *The Imperative of Responsibility* (Chicago: University of Chicago Press, 1984); J. Ellul, *The Technological Bluff* (Michigan: William B. Eerdmans, 1990). In the 1990 book Ellul continues defending the same thesis as in his former works. For a criticism of this thesis see, for example, Langdon Winner, *Autonomous Technology*, (Cambridge, MA: The MIT Press, 1977); Oberdiek, *Technology*; J. Pitt,

from supposing that technology, like science, has its own internal logic. According to this view, technology is self-determinative. It operates in accord with its own laws, independently of human dictates or will.[10] Those who defend this thesis present technology as "out of control." They say there is no human purpose driving its development, but a teleological process that begins with the wheel and passes through cloning. The autonomy conception (which incorporates the slogan of the 1933 World's Fair: "Science Finds, Industry Applies, and Man Conforms") thus provides a convenient excuse for avoiding criticism of technological development. If scientists, engineers, and politicians are not responsible for the inevitable march of technological progress, then society cannot blame them for what they cannot control.

Some scholars associate the image of autonomous technology (free from human control) with a deterministic conception of the relationship between technology and society.[11] From this point of view, technology is an independent factor, and any changes in it impose major consequences or determinants on society. Supporters of this view assume that technological progress follows a self-determined path. They believe that political, economic, and social factors can influence such progress, but these factors cannot alter the general line of development given by the "internal logic" of technology.

The most obvious problems with the thesis of technological determinism are empirical and methodological.[12] There is no way of

---

"The Autonomy of Technology," *From Artifact to Habitat*, ed. G. L. Ormiston, (Bethlehem: Lehigh University Press, 1990).

10. The metaphor of Dr. Frankenstein's monster running away from his creator and following his own will often represents the image of technology as autonomous.

11. See, for example, J. El Ellul, *Technological Society*, and J. Ellul, *The Technological System* (New York: Continuum, 1980). For a defense of the thesis of technological determinism see, for example, A. Toffler, *The Third Wave* (New York: Morrow, 1980); Merritt Roe Smith and Leo Marx, eds., *Does Technology Drive History?* (Cambridge, MA: The MIT Press, 1994).

12. For a criticism of the thesis of technological determinism see, for example, D. F. Noble, *Forces of Production: A Social History of Industrial Automatization* (New York: Alfred A. Knopf, 1984); Winner, *The Whale*; A. Feenberg, "Subversive Rationalization: Technology, Power, and Democracy", *Inquiry* 35 (1992): 301-322; N. F. Männikkö, "If a Tree Falls in the Forest: A Refutation of Technological

determining conclusively that technology is either an independent factor or the major determinant of social changes. Certainly, technology helps to shape the kind of society that we have, but from this it does not follow that technology completely and irreversibly determines our ways of living. Affirming such a thesis is to reify technology and to ignore the network of social interests that are necessary to develop and implement it. Moreover, this thesis assumes a narrow conception of technology. If we understand it not only as a result, but also as a process that includes social, cultural, economic, and political factors, then human beings can affect it.[13] In the U.S., for example, government regulations ended the use of leaded gasoline and lead-based paint after scientists discovered the serious threats that lead posed for human health. The public also played an important role in the ban of the growth-promoting drug, diethylstilbestrol (DES), used as a morning-after pill. Likewise, in Britain, concerns about ethics led to the ban of research on human embryos after 14 days of fertilization.

The doctrine of technological determinism thus blinds us to the fact that society has made choices that have controlled technology. In the second half of this century, public attitudes about technology have shifted from a positive expectation to a vocal concern about the negative or unanticipated effects of different technological achievements.[14] The damaging effects of pesticides on humans and wildlife, the health problems and deaths caused by accidents in nuclear power plants and by radioactive waste management, the dangerous side effects of hormonal contraceptives,

---

Determinism," in *Research in Philosophy of Technology* 12 (1992): 177-188.

13. See, for example, Dorothy Nelkin, *Controversy: Politics of Technical Decisions* (Beverly Hills and London: Sage, 1984)(hereafter cited as: Nelkin, *Controversy*); David Collingridge, *The Social Control of Technology* (New York: St. Martin's Press, 1980); D. MacKencie and J. Wajcman, eds., *The Social Shaping of Technology* (Philadelphia: Open University Press, 1985); W. Bijker, T. Hughes and T. Pinch, eds., *The Social Construction of Technological Systems* (Cambridge, MA: MIT Press, 1987); H. M. Collins, *Artificial Experts: Sociological Knowledge and Intelligent Machines* (Cambridge, MA: MIT Press, 1990); W. Bijker, and J. Law, *Shaping Technology-Building Society* (Cambridge, MA: MIT Press, 1992); Pierre Lemonnier, ed., *Technological Choices* (London and New York: Routledge, 1993); Stephen H. Unger, *Controlling Technology* (New York: John Wiley & Sons, 1994).

14. See, for example, D. Nelkin, "Introduction: Analyzing Risk," in *The Language of Risk*, ed., Nelkin, D. (Beverly Hills: Sage, 1985). See also Nelkin, *Controversy*.

the widespread use of certain chemicals and the threats of genetic research are some of the phenomena that have contributed to an increasing demand for public control over the kind of technology implemented in our society. Technology assessment is one of the more prominent responses to this demand.

1.2. Overview

In this chapter, I offer a brief overview of technology assessment (TA). I analyze the concept of TA and some reasons for using it. In the next section I present a succinct historical background on the development of technology assessment practice. In section four, I discuss ten components that TAs use in order to anticipate unforeseen high-order impacts. I review TA methodology in section five. I concentrate in the description of analytic, synthetic, and empirical methods used in TA. In section six, I discuss one of the techniques that dominates the decisionmaking process in TA: risk-cost-benefit analysis (RCBA). Finally, in section seven I argue that TA and RCBA presuppose philosophical judgments, such as those of aggregation, partial quantification, and importance of subjective preferences that assessors should not neglect if they desire a critical and comprehensive evaluation of technology.

1.3. Historical Background

Technology often has faced ambivalent public attitudes. Fears of innovation and desires for a "technocopia" have coexisted in western society. Certainly, technological improvements have increased the levels of comfort of many people, but they also have caused serious environmental, safety, and health problems. Cases such as the use of DDT, the commercialization of thalidomide, the nuclear accidents at Three Mile Island and Chernobyl, and so on, have generated a significant public concern about the consequences of new technologies.

In this context of increasing public awareness of unwanted technological side effects, U.S. Representative Emilio Daddario used the concept of "technology assessment" for first time.[15] In 1966, he employed the term in a report of the Subcommittee on Science, Research and Development of the House Committee on Science and Astronautics. On March 7, 1967, Daddario introduced a bill to establish a Technology Assessment Board as a vehicle to encourage discussion about the uses and consequences of science and technology. He defined TA as:

> ... a form of policy research which provides a balanced appraisal to the policymaker. Ideally, it is a system to ask the right questions and obtain the correct and timely answers. It identifies policy issues, assesses the impact of alternative courses of action and presents findings. It is a method of analysis . . . designed to uncover three types of consequences --desirable, undesirable and uncertain . . . .[16]

In July 1969, the Committee on Science and Public Policy (COSPUP) of the U.S. National Academy of Sciences (NAS)[17] produced a second report on TA. This report pointed out some deficiencies in the existing processes of governmental assessment, flaws such as overvaluing the economic benefits, ignoring externalities, or disregarding public participation. The NAS proposed some methodological, institutional, and conceptual reforms to improve the TA process. The report also recommended the creation of a government assessment organization.

---

15. For historical aspects of the concept of technology assessment see, for example, M. Kranzberg, "Historical Aspects of Technology Assessment"; and V. T. Coates, "Technology and Public Policy. The Process of Technology Assessment in the Federal Government," in *Readings in Technology Assessment* (Washington, D.C.: The George Washington University Program of Policy Studies in Science and Technology, 1975); A. L. Porter, F. A. Rossine, S. R. Carpenter, and A. T. Roper, *A Guidebook for Technology Assessment and Impact Analysis* (New York: North Holland, 1980) (hereafter cited as: Porter *et al.*, *Guidebook)*.

16. E. Q. Daddario, "Technology Assessment: Statement of Chairman, Subcommittee on Science, Research, and Development," (90th Congress, 1st Session, 1967).

17. U.S. National Academy of Sciences, *Technology: Processes of Assessment and Choice*, Documented prepared for House Committee on Science and Astronautics (Washington, D.C.: U.S. Government Printing Office, 1969).

In the summer of 1969, the Committee on Public Engineering Policy (COPEP) of the U.S. National Academy of Engineering (NAE) published a third report on TA[18]. This report was essentially a study of the methodology and potential value of assessment, as developed from three small-scale TA's in the fields of electronic teaching aids, subsonic aircraft noise, and multiphasic health screening. Based on these three exercises, the COPEP report concluded that TA is feasible and would provide useful information to Congress.

Several congressional hearings held in 1969 and 1970 culminated in a bill to create an Office of Technology Assessment. The House Committee on Science and Astronautics introduced this legislation in September 1970. A modified bill passed the Senate in September 1972, and on October 13, 1972 President Nixon signed the Technology Assessment Act of 1972 and thus created the Office of Technology Assessment (OTA) to serve Congress. Under the guidelines of the OTA an important number and variety of comprehensive TA's have been completed in such diverse areas as application of solar technology, effects of nuclear war, nutrition research alternatives and biotechnology.[19] Unfortunately, in July 1995, the U.S. Congress decided to eliminate the 22-year-old Office of Technology Assessment. This decision has left members of the Congress, as well as the public, without scientific and technical information relevant in formulating public policy on technology matters. The dissolution of this group has made it more likely that, at least in the U.S., government will not adequately evaluate various technologies. Because the OTA has been the premier agency performing TA, it is wise to investigate it further.

---

18. U.S. National Academy of Engineering, *A Study of Technology Assessment*, Document prepared for House Committee on Science and Astronautics (Washington, D.C.: U.S. Government Printing Office, 1969).

19. See, for example, U.S. Congress, Office of Technology Assessment, *Genetic Monitoring and Screening in the Workplace* (Washington, D.C.: U.S. Government Printing Office, 1990); U.S. Congress, Office of Technology Assessment, *Energy Technology Choices: Shaping Our Future* (Washington, D.C.: U.S. Government Printing Office, 1991); U.S. Congress, Office of Technology Assessment, *Biotechnology in a Global Economy* (Washington, D.C.: U.S. Government Printing Office, 1991); and U.S. Congress, Office of Technology Assessment, *Potential Environmental Impacts of Bioenergy Crop Production* (Washington, D.C.: U.S. Government Printing Office, 1993).

A Technology Assessment Board consisting of six senators and six representatives governed the OTA. The two major political parties had equal representation, and the chair alternated between the two groups. The director could initiate assessments only at the request of the Board or of congressional committees. OTA did not itself perform TA's; it initiated and directed assessments through nonprofit research groups.[20]

Although OTA represented the most visible instance of the institutionalization of TA activities, other U.S. agencies have undertaken some assessment projects. The National Science Foundation (NSF), for example, has played an important role in funding specific studies and coordinating efforts of other agencies of the federal government. It has supported TA studies to identify new technological possibilities. NSF has also instituted several studies of assessment methodology to evaluate its weaknesses, and has suggested ways in which TA's may be more effectively performed. The U.S. National Academy of Sciences (NAS) has provided much of the direction for TA. Among other contributions, it has given a list of steps that an interdisciplinary team should follow when performing evaluation of a technology.[21] The NAS has also offered a general description of risk assessment used in most of the environmental-health hazard evaluations today.[22]

Since the creation of the U.S. Office of Technology Assessment there has been a growing international interest in TA, as exhibited in several worldwide conferences on the subject.[23] International organizations such

---

20. See, for example, V. T. Coates, *A Handbook of Technology Assessment* (Washington, D.C.: The George Washington University, 1978); and Rhodry Walters, "The Office of Technology Assessment of the United States Congress: A Model For the Future?, *Government and Opposition* 27 (Winter,1992):89-108.

21. See National Academy of Sciences, Panel on Technology Assessment, *Technology: Processes of Assessment and Choice* (Washington, D.C.: U.S. Government Printing Office, 1969).

22. See National Research Council, *Risk Assessment in the Federal Government: Managing the Process* (Washington, D.C.: National Academy Press, 1983); National Research Council, *Issues in Risk Assessment* (Washington, D.C.: National Academy Press, 1993); and National Research Council, *Science and Judgment in Risk Assessment* (Washington, D.C.: National Academy Press, 1994).

23. These conferences have been held in London, 1961; Lausanne, 1972; Florence, 1975; Helsinki, 1981; and Tokyo, 1985. See V. J. McBrierty, "Technology Assessment

as the Organization for Economic Co-Operation and Development (OECD) and the European Community (EC) have taken initiatives to develop TA among their member nations.[24] The Forecasting and Assessment in Science and Technology program (FAST) reflects the contribution of the EC. The aim of FAST is twofold. On the one hand, it tries to analyze long-term changes in science and technology in order to identify new EC priorities for a common R&D policy. On the other hand, it looks for reinforcing the base for prospective thinking by means of cooperative European networks of TA's.[25]

Many European countries have conducted assessment activities. They recognize that the information generated by TA pushes political decision making to a higher level of rationality and competency and increases public awareness about technology.[26] In the former Federal Republic of Germany, two proposals (one in 1973 and another in 1977) for a national office of technology assessment failed for political reasons. The government seemed to believe that such an office would shift too much power to political parties in opposition. During the same decade, however, nonprofit groups and many large industries conducted a variety of studies dealing with impacts of nuclear power, water supply, and pollution. In the spring of 1985 the German Bundestag founded the *Enquete Kommission Technologiefolgenabschätzun*. Its task was to ascertain, by means of pilot assessments, how much evaluation work the Bundestag would be able to handle, and how a parliamentary TA organization could best be implemented. At the end of 1986, the *Kommission* advised the Bundestag to establish a small institute that would be run by members of the parliament and assisted by experts from outside.

For Parliaments At National And European Level," *Futures* 20, vol. 1 (February 1988)(hereafter cited as McBrierty, *Technology*).

24. See, for example, M. A. Boroush, K. Chen and A. N. Christakis, eds., *Technology Assessment: Creative Futures* (New York: North Holland, 1980).

25. See, for example, McBrierty, *Technology*.

26. See, for example, V. T. Coates and T. Fabian, "Technology Assessment in Europe and Japan," *Technological Forecasting and Social Change* 22 (1982): 343-361; McBrierty, *Technology*; R. Smits and J. Leyten, "Key Issues in the Institutionalization of Technology Assessment," *Futures* 20, vol. 1, (February 1988); E. J. Tuininga, "Technology Assessment in Europe," *Futures* 20, vol. 1, (February 1988).

The Netherlands Parliament faced the same kinds of political problems as the former Federal Republic of Germany. In the early 1970s, the Parliament rejected a proposal for a Dutch version of the disappeared OTA. In the early 1980s, however, it commissioned the Ministry of Education and Science to deal with the ethical and social consequences of technological development. This recommendation led to the foundation of the *Nederlandse Organisatie voor Technologisch Aspektenonderzoek* (NOTA). Although the Minister of Education established NOTA, it is an independent body consisting of a steering group with a strong scientific character and a small staff.

In France, the TA discussion began in 1976 with a government proposal to establish in the parliament a national office of technology assessment. However, until 1984 the government did not accept the proposal because it feared it would give parliament to much strength. In December of that year, the *Office Parlementaire d'Evaluation des Choix Scientifiques et Technologiques* began operations. The *Office* is a parliamentary delegation, comprised of eight *députées* and eight *sénateurs*. It has the independent rights to oversee publication of TA results and to demand inquiries into the assessment projects. Besides this parliamentary initiative, other French organizations such as the *Centre de Prospective et d'Evaluation* and the *Centre d'Etudes des Systèmes et Technologies Avancées* carry out TA-like activities.

In Sweden, interest in the social consequences of technology crystallized in the early 1970s with the development of a large number of organizations such as the Secretariat for Future Studies (SFS). Its most important task is promoting public discussion of technological innovations. A multiform group of six members determines the SFS study program after consulting different social groups.

In the United Kingdom, there was no significant or formal discussion of TA until the late 1980s. In 1989 the Congress formed the Parliamentary Office of Science and Technology (POST) to provide information on topical issues in science and technology. Initially, private resources funded the POST. Today, however, public money finances the Office.

1.4. Components of Technology Assessment

TA has evolved from an "early warning system" to a kind of policy analysis whose goal is to predict the fullest range of effects of the implementation of a technology. The most difficult and probably the most important task of a TA is not to forecast technological effects on people and the environment but to anticipate unforeseen higher-order consequences. In order to achieve its goals, TA follows a methodology that incorporates at least ten components.[27]

1.4.1. PROBLEM DEFINITION

The first component involves the specification of the scope and depth of the study and the identification of the "parties at interest." These are persons likely to gain or to lose, depending on the nature of the impact. In this phase, assessors try to answer general questions such as why to study certain technologies or projects, what are the benefits of the assessment, what are the assumptions underlying the TA formulation, and what social values merit consideration. After completing the general stage of the assessment, the analysts have to "bound" the TA. This activity allows them to proceed systematically and within the given resource constraints. The areas normally considered in bounding are the time horizons of the study, its spatial scope, institutional involvements, technologies evaluated and their range of applications, impact sectors, and policy options.[28]

1.4.2. TECHNOLOGY DESCRIPTION

This component is crucial for proper focus on impacts and policy responses. A description of a technology normally includes identification of technical parameters (i.e., efficacy, economic feasibility, and safety

27. Numerous agencies and authors have offered lists of steps to be undertaken for an interdisciplinary assessment team. For a comparison of several lists of TA components see Porter *et al.*, *Guidebook*, pp.56-57.

28. Porter *et al.*, *Guidebook*, pp. 66-73. See, also, K. Shrader-Frechette, *Science Policy, Ethics, and Economic Methodology* (Boston: Reidel, 1985), pp. 6-8 (hereafter cited as: Shrader-Frechette, *Science Policy*)

risks), alternative ways of implementing the technology, characterizing competing technologies, and definition of pertinent delivery systems. For instance, a description of a device for controlling automotive emissions would include its characterization *per se*, the scientific disciplines involved, the industries or businesses affected, the products used, and design data. It also may include an analysis of the current state of the assessed technology and of the supporting sciences. Analysts may examine influencing factors such as institutional aspects affecting development or application of the device. They may also take into account related technologies, the future state of the art, and its uses and applications under evaluation.[29]

1.4.3. TECHNOLOGY FORECAST

The goal of the technology forecast component of TA is to anticipate the future impacts of emerging technologies. At this stage, assessors can identify consequences such as future cost reductions, new applications, uncertainties, and potential breakthroughs. One might hypothesize, for example, that nuclear accidents might lead to social pressures that will, in turn, force the implementation of alternative forms of energy. Major techniques at this stage of TA are monitoring, trend extrapolation, and expert-opinion methods. Monitoring is based on the assumption that changes in the political, economic, and social environments anticipate technological developments. Trend extrapolation procedures attempt to analyze the historical progress of a technology in a mathematical or graphical model. Expert opinion methods rely on asking recognized experts in the area of assessment for their estimations.[30]

1.4.4. SOCIAL DESCIPTION

Since the core of TA is to analyze the impact of technology on society, a description of the society is crucial. This component is useful in identifying parties at interest, social values, and institutional involvements that can

---

29. See Porter *et al.*, *Guidebook*, pp. 106-109; and Shrader-Frechette, *Science Policy*, pp. 6-8.

30. See Porter *et al.*, *Guidebook*, pp. 111-130; and Shrader-Frechette, *Science Policy*, pp. 6-8.

affect, or be affected by, the implementation of a certain technology. Assessors often use descriptions at different levels. At the first level, they define the basic assumptions about society that are not likely to change in the foreseeable future. For example, they may assume that there will be neither a major war nor an important social alteration in ideology or values. A second stage of societal description deals with macroindicators such as economy, population, and education. Assessors can also describe ideological factors that have special implications for technology, such as a concern over centralization or decentralization.[31]

1.4.5. SOCIAL FORECAST

Predicting the most likely future configurations of society and projecting changes in its parameters is the purpose of the social forecast component of TA. Social forecasting is one of the least developed aspects of the assessment process, due mainly to a lack of good conceptual frameworks for the interaction between society and technology. One of the approaches that analysts tend to use for social forecasting is scenario construction. A scenario is a descriptive or chronological outline of a possible future state of society. It seeks to engender a holistic view of the pertinent social context in its relationship with the technology assessed.[32]

1.4.6. IMPACT IDENTIFICATION

The task of assessors at this stage is to identify direct and especially higher-order or indirect impacts. Direct impacts are those effects directly associated with the technology; higher-order impacts are the products of direct effects. For example, one of the unintended consequences of television may be the deterioration of social relationships because of an increasing isolation of people. Assessors may use two types of approaches to identify impacts, a reductionist strategy or a holistic one. In the reductionist approach, the analysts subdivide the totality of impacts into

31. See Porter *et al.*, *Guidebook*, pp. 135-146; and Shrader-Frechette, *Science Policy*, pp. 6-8.

32. See Porter *et al.*, *Guidebook*, pp. 146-151; and Shrader-Frechette, *Science Policy*, pp. 6-8.

smaller and more easily examined groupings. For example, they can divide them into environmental, psychological, social, economic, and legal components. Holistic strategies present the impact field as a whole, without using prestructured categories. In this case there is no division. Assessors exhibit the impacts as a continuum.[33]

1.4.7. IMPACT ANALYSIS

After identifying the impacts, assessors study the probability and magnitude of each. This investigation includes their significance, probability, costs, timing, and affected parties. Impact analysis constitutes the main body of a TA and involves a wide variety of techniques and activities such as cross-effect matrices and simulation models. Cross-effect matrices provide orderly cross-comparisons of important factors such as the influence of certain technology on other techniques, the state of society, possible alterations of environmental and social situations, or changes in policies. Both quantitative and qualitative approaches may be appropriate. Simulation models represent the dynamic behavior of a system as it varies over time. If its behavior for a given set of factors is fixed, the model is deterministic; if a number of system responses are possible, then the model is probabilistic.[34]

1.4.8. IMPACT EVALUATION

The purpose of impact evaluation is to analyze and determine the importance of identified impacts relative to the technology and to social goals. Assessors try to compare different solutions to possible consequences. They also try to assist in policy analysis. This step of the assessment involves two elements, criteria for the impact evaluation and measures for it. These elements provide the yardstick used for making judgments at this stage because they reflect the values held by the analysts or the social groups they represent. For example, criteria for impact

---

33. See Porter *et al.*, *Guidebook*, pp. 157-177; and Shrader-Frechette, *Science Policy*, pp. 6-8.

34. See Porter *et al.*, *Guidebook*, pp. 182-212; and Shrader-Frechette, *Science Policy*, pp. 6-8.

evaluation include parameters such as opportunities for investment, domestic rate of inflation, and amount of exports, all of which indicate the economic health of a country. The measures employed in the evaluation indicate the extent of satisfaction of the criteria. For instance, if the assessment is concerned about an educational technology, appropriate measures could include the technology's effects on population literacy rates.[35]

1.4.9. POLICY ANALYSIS

The objective of policy analysis is to present policy makers with a comparative study of the options available for implementing technological developments and for dealing with their desirable and undesirable consequences. It attempts to define what is likely to happen under different courses of action. Assessors conduct policy analyses (throughout the course of the TA) which may identify, for example, the economic, social, and environmental costs, risks, and benefits of implementing a particular technology. They also may consider the applicability and adaptability of the technology assessed, as well as its acceptability in relation to current laws and regulations. Analysts may or may not make explicit policy recommendations.[36]

1.4.10. COMMUNICATION OF RESUSLTS

The communication of the results of the assessment to the concerned parties is one of the most serious tasks of TA. The team of analysts decides what information and the forms of presentation it will use to present its results to the persons or groups that the technology and its evaluation are likely to affect. Achieving a useful assessment requires identifying the intended audiences and planning to meet their demands throughout the process. TA is ultimately a policy tool. Decisionmakers need to identify the advantages and disadvantages of the different

---

35. See Porter *et al.*, *Guidebook*, pp. 351-375; and Shrader-Frechette, *Science Policy*, pp. 6-8.

36. See Porter *et al.*, *Guidebook*, pp. 384-397; and Shrader-Frechette, *Science Policy*, pp. 6-8.

assessment options in order to choose the most appropriate. Thus it is important that the analysts communicate the results effectively to them. Decisionmakers also need to know the consequences of each selection and possible obstacles for the implementation of particular technological choices.[37]

Sometimes the government withholds TA results. This has been the case with information about tobacco or the effects of radiation exposure.[38] For example, between 1979 and 1981, congressional committee hearings revealed for the first time the activities of the U.S. federal government and its operatives as they implemented the testing of nuclear weapons during the 1940s and 1950s. The legislators involved in the congressional hearings accused the government of knowingly failing to give information about the dangers posed by the radioactive fallout, falsely interpreting and reporting radiation exposure rates, and giving an inaccurate estimate of the hazards. Also, at the low-level nuclear waste facility in Maxey Flats, Kentucky, the government-supervised company (Nuclear Engineering Company, NECO, later called "U.S. Ecology") illegally dumped radioactive materials offsite and refused to admit investigators to the facility once radioactive pollution had become a problem.[39] External reviewers also have criticized the U.S. Department of Energy (DOE) for communication and supervisory problems at the proposed Yucca Mountain site for high-level nuclear waste. The reviewers claim the DOE has ignored serious risks and has allegedly used Yucca Mountain as a justification for its own self- and predetermined policies. Residents of Mississippi have accused the DOE of not sharing data about their state's possible radwaste site with them. Nevadans have charged the DOE with secrecy in the face of Yucca Mountain studies, with not consulting with them, and with failure to fund Nevada's requests for research about the

---

37. See Porter *et al.*, *Guidebook*, pp. 401-417; and Shrader-Frechette, *Science Policy*, pp. 6-8.

38. See, for example, Howard Ball, *Justice Downwind,* (Oxford: Oxford University Press, 1986); Philip Fradkin, *Fallout* (Tucson: The University of Arizona Press, 1989); Richard Miller, *Under the Cloud* (New York: Free Press, 1986); and K. Shrader-Frechette, *Burying Uncertainty* (Berkeley: University of California Press, 1993), chs. 4 and 7 (hereafter cited as Shrader-Frechette, *Burying*).

39. Shrader-Frechette, *Burying*, p.71.

site.[40] However, more communication is occurring in TA as the public demands more control of technology and risk.

1.5. Technology-Assessment Methodology

The special characteristics of TA make the acceptance of a uniform methodology extremely difficult.[41] One of the main reasons for this lack of a homogeneous methodological approach is that assessors usually evaluate technologies as successful or appropriate in the context of the purpose for which they are used. For example, analysts evaluated the success of the so-called "Green Revolution" by considering not only its technical achievements, but also its objective of providing adequate food for developing nations. Also, because TA is concerned with evaluating the goals of particular technologies, it is not a purely scientific or engineering activity. TA pertains to a class of holistic studies whose objective is to incorporate all relevant information concerning a technology. Because such information may proceed from different areas, the best way to achieve reliable TAs is to include assessment team members representing different disciplines related to the project under consideration. An interdisciplinary team evaluating a particular animal growth hormone, for example, may include a biologist, a health physicist, a veterinarian, a pharmacist, an attorney, a political scientist, an ethicist, and an economist. Because of the holistic and multidisciplinary character of TA, assessment practitioners do not use a unique methodology but a battery of analytic tools and techniques specifically suited to the subject being evaluated.[42]

In general, TA methods are of three types: analytic, empirical, and synthetic.[43] Analytic procedures employ formal models that do not always depend on raw data about the external world. These types of methods are

---

40. Shrader-Frechette, *Burying*, p.71.

41. See Shrader-Frechette, *Science Policy*, pp. 12-13.

42. See Porter *et al.*, *Guidebook*, pp. 82-84, and J. F. Coates, "Technology Assessment," in *Handbook of Futures Research*, ed., J. Fowles (Westport, Connecticut: Greenwood Press, 1978).

43. See Porter *et al.*, *Guidebook*, pp. 78-87. See, also Shrader-Frechette, *Science Policy*, pp.12-13.

appropriate in dealing with well-structured problems. Analytic techniques include relevance trees, simulation models, and substitution analyses. Empirical methods use information gathering and inductively built models to explain or to calculate impacts. Assessors tend to use these second kinds of procedures in cases of well-defined problems where there is an agreement on the existence and importance of data. Techniques used in connection with empirical methods include opinion measurement, monitoring, calculation of probabilities, and trend extrapolation. Synthetic methods take advantage of the strengths of the analytical and empirical approaches. They are based on data and, at the same time, data gathering is grounded in preexisting theories. Analysts use synthetic models for more complex and ill-structured problems for which complete data are not available. Techniques such as checklists, cost-benefit analyses, cross-effect matrices, sensitivity analyses, development of scenarios, and decision analyses are associated with synthetic methods.[44]

Although assessors use each of these three approaches in connection with particular techniques and in specific problem situations, they often need many different methods and procedures to deal with the variety of problems addressed in an assessment. When evaluating a technology, there is no simple methodological prescription for which techniques to use in a certain kind of problem. While there is no menu *per se*, analysts tend to utilize checklists, relevance trees, and matrices for impact identification, while they use cost-benefit analysis and simulation modeling, for example, for impact analysis.[45]

1.6. Risk-Cost-Benefit Analysis

Among all the techniques used in TA, risk-cost-benefit analysis (RCBA) dominates the decisionmaking process. RCBA is a variant of cost-benefit analysis that incorporates notions of probability and uncertainty as a basis for determining acceptable risk decisions in technology- and environment-

44. See Porter *et al.*, *Guidebook*, pp. 78-87. See, also Shrader-Frechette, *Science Policy*, pp.12-13.

45. See Porter et al., *Guidebook*, pp. 78-87. See, also Shrader-Frechette, *Science Policy*, pp.12-13.

related problems. The objective of RCBA is to discover whether the benefits derived from a proposed project outweigh the risks and costs incurred. Although the use of RCBA to analyze the economic impacts of a specific project or technology has its origins in the investment decisions of private firms, currently most U.S. regulatory agencies use RCBA to help determine their technological policies.[46]

RCBA incorporates a broad collection of techniques of economic evaluation in a common framework. Most RCBAs are structured according to four main steps. The first step consists of the definition of the subject of analysis, including the parties at interest. This is the most important stage in a RCBA because it provides limits and direction for the entire study. Subject definition incorporates a detailed technical description of the topic of assessment, the period of development of the analysis, background assumptions relating to the technical and social context in which the development will occur, and the parties who will pay the costs and receive the benefits.[47]

The goal of the second step of RCBA is the identification and description of the anticipated risks, costs and benefits. Such an identification must address not only direct monetary factors, but also the economic aspects of other impacts originated with the new project (e.g., the dollar value of recreational use of a tourist village developed around a new road).[48]

In the third step of RCBA, assessors usually measure direct costs and benefits, externalities, and higher-order impacts. Direct costs are the actual

---

46. See Shrader-Frechette, *Science Policy*, pp. 14-17. See also, U.S. National Institute of Health, *Cost Savings Resulting from NIH Research Support* (Washington D.C.: U.S. Department of Health and Human Service, 1990); Committee on Science, Space, and Technology, *Risk Assessment: Strengths and Limitations of Utilization for Policy Decisions* (Washington, D.C.: U.S. Government Printing Office, 1991); Committee on Science, Space, and Technology, *Strengthening Risk Assessment within EPA* (Washington, D.C.: U.S. Government Printing Office, 1994); Committee on Science, Space, and Technology, *Risk Assessment Research* (Washington, D.C.: U.S. Government Printing Office, 1994).

47. See Porter *et al.*, *Guidebook*, pp. 255-57; and Shrader-Frechette, *Science Policy*, p. 16.

48. See Porter *et al.*, *Guidebook*, pp. 257-58; and Shrader-Frechette, *Science Policy*, p. 16.

charges invested in a technological project (e.g., the price of the land for building a new airport). Direct benefits are the gains received for the project (e.g., in the case of an airport, payment for using the new facilities). Externalities are the social risks, costs, and benefits carried by the public for which they do not receive compensation or charge. A positive externality of the airport, for example, is the job opportunities it originates. Negative externalities are the air and noise pollution caused by airplanes and vehicular traffic. Higher-order impacts are indirect and delayed risks, costs, and benefits arising from the effect of direct or foreseen consequences.[49]

Finally, the last step of RCBA consists of an evaluation of the various economic impacts. At this stage analysts usually select decision criteria and a discount rate. They also actually perform the economic study, including sensitivity analysis. Discounting is the means by which one compares current dollar investments with future returns. The discount rate measures how much more value resources used now have over resources employed in the future. For example, at a 6% discount rate, a person who invests $100 now would expect $112.36 in two years. At the same 6% discount rate, $100 in 20 years would be worth $31 today. Some of the most important decision criteria for evaluating economic impacts are net present value, annual equivalent, internal rate of return, payback period, and composite economic and factor profiles. The net present value method consists of calculating risks, costs, and benefits and the net benefits for each year of life and discounting them back to the present. In annual equivalent calculation, analysts compute present worth and then convert it to an equivalent stream of annual amounts. Internal rate of return is the discount rate at which the present worth of a series of net benefits is equal to the present worth of a series of net costs and risks. Payback period is the time required for a project to recover its costs. Composite economic and factor profiles weigh not only quantitative elements but also qualitative aspects of the project. Analysts must consider the nonmonetary components against the monetary terms.[50]

---

49. See Porter *et al.*, *Guidebook*, pp. 264-69; and Shrader-Frechette, *Science Policy*, p. 16.

50. See Porter *et al.*, *Guidebook*, pp. 269-78; and Shrader-Frechette, *Science Policy*, p. 17.

Another important method for evaluating economic impacts is sensitivity analysis.[51] In using this procedure, assessors alter the risk, cost, or benefit variables, their interrelationships, or their values, to examine how such modifications affect the values and range of impacts. By doing sensitivity analysis, assessors hope to discover how dependent their conclusions are on their assumptions and on the precision of their risk-cost-benefit data. They also try to identify the leverage points on which to base strategies for modifying their results. For example, after the evaluation of a new airport appears advantageous terms of risk-cost-benefit analysis, assessors may be concerned with some uncertain conditions regarding the number of jobs the facility will create. In order to compensate for such uncertainties, the analysts can decide to modify some of the parameters of the assessment. Thus, they can explore how such changes affect the values and consequences of their evaluation. For instance, they can decide to modify the location of the airport, perhaps moving it to a more populated area. Finally, they can evaluate the modifications in the data that the initial assessment offered and corroborate whether the change of location would increase or decrease the number of jobs.[52]

1.7. Technology Assessment Presupposes Philosophical Judgments

Analysts often have argued that TA and RCBA must be completely neutral and objective.[53] Numerous engineers and assessors have claimed that

---

51. See Shrader-Frechette, *Science Policy.*, p. 17; Porter *et al.*, *Guidebook*, pp. 208-09; Department of Veterans Affairs, *Cost-Benefit Analysis Handbook* (Washington, D.C.: Assistant Secretary for Finance and Planning, 1989). See also, E. M. Gramlich, *A Guide to Benefit-Cost Analysis* (Englewood Cliffs, NJ: Prentice Hall, 1990); P. O. Johansson, *Cost-Benefit Analysis of Environmental Change* (Cambridge: Cambridge University Press, 1993).

52. See Porter *et al.*, *Guidebook*, pp. 208-09; Shrader-Frechette, *Science Policy*, p. 17.

53. See, for example, U.S. Congress, Office of Technology Assessment, *Annual Report to the Congress for 1976* (Washington, D.C.: U.S. Government Printing Office, 1976), p. 4; U.S. Congress, Office of Technology Assessment, *Annual Report to the Congress for 1978* (Washington, D.C.: U.S. Government Printing Office, 1978), p. 7.

technology assessment deals with facts, not with values. Therefore, they say that TA ought to follow a principle of "complete neutrality" and objective description of technical facts and ought not to include any normative or evaluative components.

Although it is obviously necessary to protect technology assessment from intentional bias, proponents of the principle of complete neutrality typically confuse different kinds of values, not all of which are equally objectionable. Supporters of the value-free assumption often mistake bias values, constitutive values, and contextual values.[54] Bias values are deliberate misinterpretations and omissions to serve one's own interests. Falsification of data, manipulation of statistics or interpretations of data to support one's own prejudices are examples of bias values. Cases of scientific fraud constitute clear representations of these kinds of values.[55] Constitutive values are the source of the rules determining acceptable scientific method or practice. What counts as evidence for a theory or as an explanation of a phenomenon depends on certain assumptions about scientific method. Some examples of constitutive values are the empirical adequacy of the theories, their consistency with accepted theories in other domains, objectivity, and simplicity.[56] Finally, contextual values are social, personal, and cultural preferences. They have an impact on the kind of scientific research pursued and the method employed. Contextual values (such as the profit motive) also can motivate the acceptance of global assumptions that decide the character of research in an entire field. The research on human interferon constitutes an example of the influence of contextual values.[57] Cultural and economic preferences have strongly

---

54. See H. Longino, *Science as Social Knowledge* (Princeton, NJ: Princeton University Press, 1990) (hereafter cited as Longino, *Science*).

55. See, for example, S. J. Gould, *The Mismeasure of the Man* (New York: Norton, 1981); Committee on Government Operations, *Are Scientific Misconduct and Conflicts of Interest Hazardous to our Health?* (Washington, D.C.: U.S. Government Printing Office, 1990); Committee on Energy and Commerce, *Scientific Fraud* (Washington, D.C.: U.S. Government Printing Office, 1992).

56. See, for example, C. G. Hempel, *Philosophy of Natural Science* (Englewood Cliffs, NJ: Prentice Hall, 1966); K. Popper, *The Logic of Scientific Discovery* (London: Hutchinson, 1959); T. Kuhn, *The Structure of Scientific Revolutions* (Chicago: University of Chicago Press, 1962, 1970)(hereafter cited as Kuhn, *Structure*).

57. See Longino, *Science*, pp.86-89. See, also, H. Longino, "Biological Effects of

influenced industrial microbiology, as practiced by small firms founded by biochemists. These values have determined how scientists test and announce research results. Thus because of interferon's alleged therapeutic significance in treating cancer, profit motives heavily influenced the scientific activity surrounding the research.

While studies of science and technology can and should avoid bias values, it is in principle impossible to elude constitutive values, and it is in practice impossible to avoid contextual values for the following reasons.[58] First, all investigation is unalterably theory-laden in requiring both a definition of the research problem and a criterion for relevant evidence.[59] Use of evaluative definitions and criteria mean that the investigation can never be neutral in the sense of avoiding constitutive values. Second, it is in practice impossible to avoid contextual values because they often fill the gap resulting from inadequate knowledge. Because incomplete information limits any research, ignorance about a particular phenomenon provides an opportunity for the determination of information using contextual (social and moral) values.[60] An example of the mediating influence of these values is the use of differential distribution of hormones between males and females to explain behavior differences between the sexes.[61] Because scientists have observed that testosterone produces aggressive behavior in laboratory animals, many scholars have inferred from this observation the theory that testosterone can explain male-female differences.[62] Frequently, however, filling the gap between the observation of laboratory animals and the theory about humans is the contextual norm that aggressiveness is a feature of male rather than female behavior or that it is biologically determined in humans.[63]

Low Level Radiation: Values, Dose-Response Models, Risk Estimates," *Syntheses* 81 (1990): 391-404.

58. See Longino, *Science*, ch. 1; and Shrader-Frechette, *Science Policy*, pp. 67-72.

59. See, for example, Norwood Russell Hanson, *Patterns of Discovery* (Cambridge: Cambridge University Press, 1958); Michael Polanyi, *Personal Knowledge* (New York: Harper and Row, 1958); Kuhn, *Structure*.

60. See Longino, *Science*, ch. 1; Shrader-Frechette, *Science Policy*, pp. 71-72.

61. Longino, *Science*, ch. 7.

62. Longino, *Science*, ch. 7.

63. Longino, *Science*, ch. 7.

TA is not solely about facts--assuming it is possible, in any way, to make a clear distinction between facts and values in the assessment of technologies.[64] On the contrary, TA--and one of its preeminent methods, RCBA--presuppose a number of methodological, ontological, and ethical judgments that need critical evaluation if we wish to improve them. Basic methodological assumptions of RCBA, such as those of aggregation and quantification, reveal the value-laden nature of risk-cost-benefit methodology. For example, when assessors use RCBA in the evaluation of a technology, they quantify and combine different kinds of parameters. In the assessment of a new airport facility, analysts have to aggregate measures of many factors such as noise pollution, traffic congestion, effects on property values, satisfaction of the neighbors, number of jobs created, accessibility, and economic benefits. The result is an index that is supposed to provide a meaningful measure of the total costs and benefits of the facility.[65] However, the addition of dissimilar items does not always yield a meaningful value. Factors such as noise pollution and accessibility have elements that it may make no sense to add. On the one hand, assessors can measure noise in decibel level, persistency, and time of the day; on the other, these elements in different combinations could vary broadly in their perturbing effects. In contrast, accessibility involves measures of variables with dissimilar values for people of different ages, with diverse incomes, or at various distances from the airport. To combine them into some unique number would be meaningless.

Another problem with aggregation is that adding measurable items, such as monetary costs or benefits, does not give knowledge about

---

64. For an analysis of the difficulty --and uselessness-- of the distinction between facts and values in the assessment of technology, see A. C. Michalos, "Technology Assessment, Facts, and Values," in *Philosophy and Technology*, eds., P. T. Durbin and F. Rapp (Dordrecht, Holland: Reidel, 1983), pp. 59-81.

65. For examples of the use of indicators in different disciplines see, Sally Baldwin, Christine Godfrey, and Carol Propper, *Quality of Life. Perspectives and Policies* (London and New York: Routledge, 1990); U.S. Environmental Protection Agency, *Biological Populations as Indicators of Environmental Change* (Washington D.C.: Office of Policy, Planing, and Evaluation, 1992); Center for Health Economics Research, *Access to Health Care. Key Indicators for Policy* (Princeton, NJ: The Robert Wood Johnson Foundation, 1993); R. Mark Rogers, *Handbook of Key Economic Indicators* (Burr Ridge, IL: Irving, 1994).

nonmeasurable welfare. Often, money does not act as a real indicator of welfare. Uncritical acceptance of an aggregate indicator means that assessors' conclusions typically do not take into account ethical notions, perhaps ignoring the fact that market prices do not equal the full value of things. As a consequence of uncritical use of the aggregation assumption, assessments frequently ignore critical variables, such as the distributional effects of risks, costs, and benefits, in favor of the *status quo*.[66]

No less important in RCBA is the assumption of partial quantification. This assumption is that quantitative values ought not to be placed on qualitative, subjective, or nonmarket risks, costs, and benefits, such as the aesthetic impacts of technology. However, failure to quantify nonmarket values may lead to a number of undesirable consequences. If assessors do not quantify all factors, people can ignore crucial RCBA considerations, making it impossible to determine the actual desirability of different technological programs and policies. They can neglect consideration of political and social solutions to pressing technical problems. Thus, not quantifying certain factors may cause analysts to ignore them. When assessors disregard qualitative social values, while emphasizing quantitative parameters, they tend to produce a pro-technology bias among policymakers. This is historically documented: failure to consider the nonmarket costs of increased cancers and genetic injuries promoted the conclusion that the computed tomography (CT) scanner was risk-cost-beneficial. But, because assessors did not quantify the hazards posed by cancer and genetic damage, their evaluation of the CT scanner did not accurately identify its risks, costs, and benefits.[67]

In the same way, ontological assumptions such as taking market models as the correct vehicles for expressing all risks, costs, and benefits, or taking subjective preferences as the measure of the value of certain risks, clearly shape the evaluation of technology. In accepting the market

---

66. For a critical analysis of aggregate indexes see, for example, Shrader-Frechette, *Science Policy*, pp. 121-146; Judith Eleanor Innes, *Knowledge and Public Policy. The Search for Meaningful Indicators* (New Brunswick, NJ: Transaction, 1990); Cass R. Sustein, "Well-Being and the State," *Harvard Law Review* 107 (April 1994): 1303-1327; Sudhir Anand and Christopher J, Harris, "Choosing a Welfare Indicator," *The American Economic Review* 84 (May 1994): 226-231.

67. See Shrader-Frechette, *Science Policy*, pp. 152-200.

system for calculating risks, costs, and benefits, assessors fail, for example, to take into account social costs or externalities. Therefore, analysts support assessment conclusions based on incomplete evaluations of risks, costs, and benefits. Also, using subjective feelings as criteria for assessing well-being, analysts tend to promote technological choices that meet citizen's demands rather than their needs.[68]

Despite the philosophical complexities and problems associated with TA and RCBA, their use is growing around the world. Therefore, critical evaluation of their presuppositions is important because such assumptions have ethical and political consequences for the decisionmaking process. For example, assuming that an aggregate index of measurable items gives us knowledge about nonmeasurable factors may sanction policies that promote an unequal distribution of risks, costs and benefits or it may favor industrial, as opposed to public, preferences about technology.

1.8. Summary and Conclusion

The main goal of technology assessment is to discover and foresee the beneficial and harmful risks and consequences of various electronic, chemical, agricultural, industrial, biological, or medical technologies, and therefore to provide a sound basis for developing public policy regarding them. Perhaps the most significant contribution of technology evaluation is that it can offer a systematic framework for analyzing new procedures prior to their implementation by industry and before government regulation. Without such evaluation, institutionalized through government offices, industry would likely introduce new technologies without a rigorous investigation of their social, ethical, and legal impacts.

In this chapter I have offered a brief overview of technology assessment and have analyzed the concept of TA and some reasons for using it. I have presented a succinct history of the development of technology assessment and have discussed ten components that TAs use in order to anticipate unforeseen higher-order impacts. After reviewing TA methodology, I examined one of the techniques that dominates that

68. See Shrader-Frechette, *Science Policy*, pp. 123-144.

dominates the decisionmaking process in TA: risk-cost-benefit analysis (RCBA). Finally, I have argued that TA and RCBA presuppose philosophical judgments, such as those about aggregation, partial quantification and subjective preference judgments that assessors should not neglect if they desire a critical and comprehensive evaluation of technology.

Because of the importance of technology assessments for grounding sound public policies, analysis of their ethical and epistemological assumptions is also essential. Inadequate technology assessments may be as dangerous as no assessment at all, because they seem to legitimize particular policy decisions with expert testimony. In the following chapters, I show how problematic evaluations of assisted-conception techniques, such as *in vitro* fertilization (IVF), may have wrongly guided public policies concerning infertility treatments. Questionable technology assessments of IVF may have contributed to endanger women's well being. It is also responsible for the emergence of ethical problems concerning risks to women's health, the common good, and rights to free informed consent.

CHAPTER 3

IN VITRO FERTILIZATION: CONTEXTUALIZING THE TECHNOLOGY

1.1. Introduction

Approximately 23,000 children have been born through *in vitro* fertilization (IVF) in the U.S..[1] Data on the cost of infertility services shows total public and private expenditures over $2 billion in 1994 in this country.[2] The number of fertility clinics has increased from 30 in 1985 to over 200 in 1997.[3] As these data indicate, the area of assisted-conception technologies is an intensively developed field in the industrialized nations. The introduction of these new procedures in our society has raised a number of ethical, social, and legal issues. In this chapter my main concern is with ethical problems related to the clinical utilization of IVF as a medical technology for infertility treatment. Although important, I do not analyze here issues of embryo experimentation. Instead, my main concern is with the technology itself and with the context in which it has developed. Like other technologies, faulty evaluations and assessments have characterized IVF. This analyses attempts to help correct such problems.

---

1. See J. Palca, "A Word to the Wise," in *The Hastings Center Report*, 24 (March-April 1994): 5 (hereafter cited as Palca, Word).

2. S. Brownlee, "The Baby Chase: Millions of Couples Have Infertility Problems, and Many Try High-Tech Remedies. But Who Minds the Price Clinics They Turn to?" in *News & World Report* 117:22 (1994): 84-90 (hereafter cited as Brownlee, Baby Chase).

3. See Centers for Disease Control and Prevention (CDC), *Assisted Reproductive Technology Success Rate in the United States: 1995 National Summary and Fertility Clinic Report* (Atlanta, GA: Government Printing Office, 1997) (hereafter cited as CDC, *Success Rate*).

1.2. Overview

In order to contextualize the issues of biomedical ethics and technology assessment that I analyze in this work, I start this chapter with a description of *in vitro* fertilization and embryo-transfer, or embryo-replacement (IVF-ET, ER) procedure, usually abbreviated as IVF. (Initially, specialists used the term `embryo transfer' to describe the insertion of an embryo into the woman's uterus. Later, however, professionals started to employ this term to describe the implantation of the embryo in another woman. To avoid confusion I shall use 'embryo transfer' to refer to placing the embryo into the uterus of the eggs' donor. In case of transfer into a different woman I shall use `embryo donation.') In section four I briefly describe the scientific development of IVF. In section five, I give an account of some ethical dilemmas that clinical IVF raises in our society. Finally, in section six I summarize some laws and regulations related to IVF and associated techniques in the United States, Australia, Canada, and several western European countries.

1.3. In Vitro Fertilization: The Medical Technique

According to the National Center of Health Statistics and the World Health Organization (WHO) between 8 and 10 per cent of couples in the industrialized countries have reproductive problems.[4] They experience difficulties in conceiving children. The factors that most often contribute to fertility disorders among women are problems in ovulation, blocked or scarred fallopian tubes, and endometriosis (the presence in the lower abdomen of tissue from the uterine lining). In about half of the couples with reproductive problems, there is a contributing male factor. Among men, most cases of infertility are a consequence of abnormal or too few sperm. About 1.6 per cent of couples seeking infertility treatment in

---

4. See, Office of Technology Assessment, *Infertility: Medical and Social Choices* (Washington, D.C.: U.S. Government Printing Office, 1988) (hereafter cited as OTA, *Infertility*); W. D. Mosher and W. F. Pratt, *Fecundity and Infertility in the United States, 1965-1988* (Hyattsville, Md.: National Center for Health Statistics, 1990); and World Health Organization, *Recent Advances in Medically Assisted Conception* (Geneve: WHO, 1992).

industrialized nations decides to use IVF.[5] In order to understand fully the epistemological, social, legal, and ethical issues that arise with the evaluation of this technology, I shall first offer a description of the procedure.

For clarification, before describing briefly the different IVF steps, I shall define the terminology used in this chapter to refer to the female egg before and after fertilization. 'Oocyte' is the female gamete or ovum formed in the ovary. The 'zygote' is the fertilized oocyte formed by the fusion of egg and at least one sperm, containing DNA from both. The 'blastocyst' is a fluid-filled sphere developed from a zygote. The embryo develops from a small cluster of cells within the sphere. After this terminological clarification, I now describe the various steps of IVF.

In its most basic case (i.e., the woman undergoing IVF provides her own eggs, and her husband or partner supplies the sperm) the technique of IVF consists of several stages. First, doctors stimulate the woman's ovaries with different hormones to produce multiple oocytes. Next, they remove the eggs from her ovaries through procedures such as laparoscopy or ultrasound-guided oocyte retrieval. After preparation of semen, specialists fertilize the mature eggs in a laboratory dish with the husband's or partner's sperm. If one or more normal looking embryos result, specialists place them (normally between three and five) in the woman's womb to enable implantation and possible pregnancy.

What follows is a more detailed description of the steps involved in *in vitro* fertilization with the embryo-transfer technique. I shall use here again the simplest case of IVF—in which the gametes come from the woman undergoing treatment and from her partner.

1.3.1. OVARIAN STIMULATION

Professionals normally assume that there is a greater chance of conception if they transfer more than one embryo into the uterus. In order to generate them, doctors need to collect and fertilize a certain number of oocytes. Because women usually produce only one oocyte per menstrual cycle, practitioners give them estrogens, gonadotropins, and progesterone hormones that stimulate the growth of multiple ovarian

---

5. Subcommittee on Health and The Environment, *Fertility Clinic Services* (Washington, D.C.: Government Printing Office, 1992), p. 16 (hereafter cited as SHE, *Fertility*).

follicles. Ovarian induction regimes in current use include clomiphene citrate (CC), human menopausal gonadotropin (hMG), purified follicle-stimulating hormone (FSH), human chorionic gonadotropin (hCG), and gonadotropin-releasing hormone (GnRH). These hormone treatments induce the maturation of an average of 7-8 eggs per menstrual cycle.[6]

1.3.2. OOCYTE RETRIEVAL

The aim of the second step of IVF, oocyte retrieval, is to collect the woman's eggs. Professionals generally monitor stimulated ovarian cycles to ensure the efficacy and safety of therapy. Doctors examine the level of estrogen, progesterone, and luteinizing hormone (LH) in the woman's blood and urine. The rise in LH indicates that ovulation is imminent. They also may use ultrasound in order to ascertain the number and diameter of growing follicles. When the follicle is 17-18 mm. in diameter, the specialist knows that ovulation is impending.[7]

Doctors schedule the recovery of oocytes after the eggs' maturation but prior to ovulation. The two main techniques of oocyte retrieval are laparoscopy and ultrasonically-guided procedures. Laparoscopic follicular aspiration was the dominant method of recovery for several years. Laparoscopy consists of making two or three small incisions in the abdominal wall to allow the insertion of the laparoscope, (a slender

---

6. See M. C. Macnamee and P. R. Brinsden, "Superovulation Strategies in Assisted Conception," in *A Textbook of in Vitro Fertilization and Assisted Reproductive Technology*, eds., P. R. Brinsden, and P. A. Rainsbury (Park Ridge, NJ: The Parthenon Publishing Group, 1992), pp. 111-126; I. Calderon and D. Healy, "Endocrinology of IVF," in *Handbook of In Vitro Fertilization*, eds. A. Trounson and D. K. Gardner (Boca Raton: CRC Press, 1993), pp. 1-16; M. P. Steinkampf and R. E. Blackwell, "Ovulation Induction," in *Textbook of Reproductive Medicine*, eds., B. R. Carr and R. E. Blackwell (Norfolk: Appleton & Lange, 1993), pp. 469-480; E. E. Wallach, "Induction of Ovulation: General Concepts," in *Reproductive Medicine and Surgery*, eds., E. E. Wallach and H. A. Zacur (St. Louis: Mosby, 1994), pp. 555-568; E. Y. Adashi, "Clomiphene Citrate-Initiated Ovulation: The State of the Art," in *Reproductive Medicine and Surgery*, eds., Wallach and Zacur, pp. 593-610; B. Lunenfeld and V. Insler, "Human Gonadotropins," in *Reproductive Medicine and Surgery*, eds., Wallach and Zacur, pp. 611-638; and H. A. Zacur and Y. R. Smith, "Gonadotropin-Releasing Hormone and Analogues in Ovulation Induction," in *Reproductive Medicine and Surgery*, eds., Wallach and Zacur, pp. 639-648.

7. See, for example, R. J. Stillman and D. I. Arbit, "Monitoring of Ovulation," in *Reproductive Medicine and Surgery*, eds., Wallach and Zacur, pp. 569-591.

optical device). It requires general anesthesia and distension of the abdomen with carbon-dioxide gas.[8]

Professionals developed ultrasound-directed techniques for use in cases where the ovaries were inaccessible to the laparoscope. As equipment has improved and operators have developed greater expertise, in most cases ultrasound-guided procedures have replaced laparoscopy.[9] Ultrasonically-guided oocyte procedures are safer and cheaper than laparoscopy because they avoid the dangers of general anesthesia and hospitalization. Practitioners may perform them through a transabdominal approach or through a recently developed transvaginal route. In the transabdominal technique, doctors, guided by the ultrasound, insert a catheter through the abdomen and through a full urinary bladder to reach the ovary. In the transvaginal method, medical professionals insert a catheter through the vagina, which allows them to get closer to the ovary than does the prior procedure. The transvaginal method is also less painful.[10]

1.3.3. SEMEN PREPARATION

Shortly after oocyte recovery, doctors obtain a sperm sample from the male partner. Semen preparation, the third step of IVF, is of major importance. Through a process of liquefication and centrifugation,

---

8. See R. Hunt, "Operative Laparoscopy: Patient Selection and Instrumentation," in M. M. Seibel, *et al.*, eds., *Technology and Infertility* (New York: Springer-Verlag, 1993), pp.51-60; P. R. Brinsden, "Oocyte Recovery and Embryo Transfer Techniques for in Vitro Fertilization," in *A Textbook of in Vitro Fertilization and Assisted Reproductive Technology*, eds., Brinsden, and Rainsbury, pp. 139-153; P. J. Taylor and J. V. Kredentser, "Diagnostic and Therapeutic Laparoscopy and Hysteroscopy and Their Relationship to in Vitro Fertilization," in *A Textbook of in Vitro Fertilization and Assisted Reproductive Technology*, eds., Brinsden and Rainsbury, pp. 73-92 (hereafter cited as Taylor and Kredentser, *Diagnostic*); See O. K. Davis and Z. Rosenwaks, "Assisted Reproductive Technology," in B. R. Carr and R. E. Blackwell, *Textbook of Reproductive Medicine*, (Norfolk: Appleton & Lange, 1993) pp. 571-586 (hereafter cited as: Davis and Rosenwaks, *ART*).

9. Davis and Rosenwaks, *ART*.

10. See Taylor and Kredentser, *Diagnostic*; Brinsden, *Oocyte*; O. K. Davis and Z. Rosenwaks, *ART*.

specialists separate the sperm from the seminal plasma because the latter contain factors that inhibit the sperm's ability to fertilize the egg.[11]

1.3.4. IN VITRO FERTILIZATION

After preparation of semen, professionals inseminate the oocytes with roughly 150,000 motile sperm per egg. Next, they incubate the sperm and oocytes in culture dishes for 12 to 18 hours. Practitioners monitor the eggs to confirm fertilization. If this occurs, they transfer the fertilized ova or zygote to a growth medium and incubate them for another 24 to 60 hr. Prior to transfer into the woman's uterus, doctors assess embryo quality on the basis of morphological criteria (symmetry and shape). If there are more than four or five normal embryos, specialists cryopreserve them for subsequent cycles in case implantation does not occur after initial transfer.[12]

1.3.5. EMBRYO TRANSFER

The transfer of the embryos is normally a nonsurgical technique. Most of the time when women undergo IVF treatment, doctors transfer more than one in order to increase the probability of implantation and pregnancy. Practitioners load the embryos into a catheter and introduce them into the woman's uterus. Before the procedure, doctors position the woman on a suitable examining table with a sterile speculum placed in the vaginal canal. Next, they place the catheter (with the embryos) through the cervix into the uterine cavity. Finally, they inject them with a small syringe attached to the catheter. Specialists recommend that the woman rest for some time after the transfer.[13]

---

11. See, H. W. G. Baker, F. L. H. Ng, and D. Y. Liu, "Preparation and Analysis of Semen for IVF/GIFT," in *Handbook of In Vitro Fertilization*, eds., in Trounson and Gardner, pp. 33-56.

12. See A. Trounson and J. Osbon, "In Vitro Fertilization and Embryo Development," in Trounson and Gardner (eds.), *Handbook of In Vitro Fertilization*, pp. 57-83; D. K. Gardner and M. Lane, "Embryo Culture Systems," in eds., *Handbook of In Vitro Fertilization*, eds., Trounson and Gardner, pp. 85-115; Davis and Rosenwaks, *ART*; and M. D. Damewood, "In Vitro Fertilization and Assisted Reproductive Technologies," in *Reproductive Medicine and Surgery*, eds., Wallach and Zacur, pp. 845-859 (hereafter cited as Damewood, In Vitro).

13. See Davis and Rosenwaks, *ART*; and Damewood, In Vitro.

1.4. In Vitro Fertilization: The Scientific History

Although thousands of children have been born worldwide through IVF, and it is increasingly common in infertility practice,[14] researchers did not originally devise the technique as a solution to reproductive problems. IVF research was important to the animal breeding industry for the study of fertility, development, and heredity. Research on IVF started to focus on the alleviation of infertility only in the 1960s.[15]

In 1890, Cambridge physiologist Walter Heape carried out the successful transfer of embryos in rabbits for the first time.[16] His purpose was to investigate the belief that the effects on the uterus of bearing a certain set of offspring were transmissible to later progeny from the same animal. Such investigation was of great interest to animal breeders, who thought that superior females could be corrupted by unintended coupling with an inferior male. Heape removed two embryos from an Angora doe rabbit that had been mated with an Angora butch. Next, he placed the

---

14. Society for Assisted Reproductive Technology, American Fertility Society, "Assisted Reproductive Technology in the United States and Canada: 1991 Results from the Society for Assisted Reproductive Technology Generated from the American Fertility Society Registry," *Fertility and Sterility* 59:5 (1993): 956-962.

15. E. Yoxen, *Unnatural Selection?* (London: Mackays, 1986), ch. 4; and M. C. Chang, "Acceptance of the Award of Scientific Merit," in *In Vitro Fertilization and Other Assisted Reproduction*, eds., H. W. Jones, Jr. and C. Schrader (New York: The New York Academy of Sciences, 1988), pp. xvii-xviii. For historical overviews of the assisted-conception technologies see, for example, A. Westmore, "History," in C. Wood and A. Trounson, eds., *Clinical In Vitro Fertilization* (Berlin: Springer, 1984), pp. 1-10; S. Fishel, "IVF -- Historical Perspective," in *In Vitro Fertilization: Past, Present, and Future*, eds., S. Fishel and E. M. Symonds (Oxford: IRL Press, 1986), pp. 1-16; L. M. Talbert, "The Assisted Reproductive Technologies. An Historical Overview," *Archives of Pathology and Laboratory Medicine* 116:4 (1992): 320-322; S. H. Chen and E. E. Wallach, "Five Decades of Progress in Management of the Infertile Couple," *Fertility and Sterility* 62:4 (1994): 665-685 (hereafter cited as Chen and Wallach, Five); and N. Perone, "In Vitro Fertilization and Embryo Transfer. A Historical Perspective, *Journal of Reproductive Medicine* 39:9 (1994): 695-700 (hereafter cited as Perone, Transfer).

16. W. Heape, "Preliminary Note on the Transplantation and Growth of Mammalian Ova within a Uterine Foster-Mother," *Proceedings of the Royal Society* 48 (1891): pp. 457-458 (hereafter cited as Heape, Preliminary); and W. Heape, "Further Note on the Transplantation and Growth of Mammalian Ova within a Uterine Foster-Mother," *Proceedings of the Royal Society* 62, (1897/98: 178-183 (hereafter cited as: Heape, Further Note).

embryos into the Fallopian tube of a Belgian Hare doe rabbit mated with a butch of its species. The Belgian Hare gave birth to six rabbits, four of which resembled her and her mate, and two were Angoras. Heape determined that the experiment proved the feasibility of the embryo transfer technique. It also showed no evidence that the uterus of the foster-mother had affected the Angora rabbits.[17]

In 1930 the American biologist, Gregory Pincus, started to work on IVF and embryo replacement in rabbits. In 1934 he and his colleague E. Enzmann claimed to have fertilized one rabbit egg *in vitro.*[18] Some controversy exists over the results of his studies. Critics have suggested that Pincus did not actually achieve the egg fertilization, but that it was quite possible for the egg to have divided spontaneously.[19] Pincus' work, however, laid the foundation for the stages of oocyte maturation and fertilization of mammalian eggs. He discovered that oocytes removed from their follicles would continue their maturation. He obtained oocytes for experimental purposes through surgery on the ovary. This allowed a easier observation of their maturation and fertilization.[20]

From 1938 to 1952, Harvard physician John Rock and his collaborator Arthur Hertig examined over a thousand fallopian tubes and wombs from women who had hysterectomies. They timed the surgeries to occur just after ovulation, hoping that if the women had had intercourse in the days before ovulation, fertilization could result. In 14 years Roch and Hertig collected 34 fertilized eggs from 210 unknowingly pregnant women, making possible the study of early human embryos.[21] (Issues of free informed consent were simply disregarded here.)

John Roch was also interested in attempting human *in vitro* fertilization. With his assistant, Miriam Menking, he collected a number of unfertilized eggs from women undergoing hysterectomies. Next, they exposed the ova to the sperm of interns paid by Rock. In 1944, they

---

17. Heape, Preliminary; and Heape, Further Note.

18. G. Pincus and E. V. Enzmann, "Can Mammalian Eggs Undergo Normal Development In Vitro?" *Proceedings of the USA Natural Academy of Sciences* 20(1934): 121 (hereafter cited as Pincus and Enzmann, Mammalian).

19. See Perone, *Transfer*, p. 695.

20. See Pincus and Enzmann, Mammalian.

21. A. T. Hertig *et al.*, "Thirty-four Fertilized Human Ova, Good, Bad, and Indifferent, Recovered from 210 Women of Known Fertility: a Study of Biologic Wastage in Early Human Pregnancy," *Paediatrics* 23 (1959): 202-211.

allegedly achieved the fertilization of an egg.[22] However, their claim is not now generally accepted because of faulty research methodology.

Studies on agricultural and laboratory animals introduced the idea of transplanting embryos in the 1940s and early 1950s. But these were embryos created *in vivo* by normal mating or by artificial insemination and then flushed out of the animal's body and replaced in another. In 1959, however, M.C. Chang of the Worcester Foundation in Massachusetts reported successful *in vitro* fertilization in rabbits. For his experiment, Chang took sperm from male rabbits with specific traits not present in the female ovum donor. The presence of the male traits in the offspring provided unequivocal proof that the sperm had transmitted genetic information to the young.[23]

The prospect of understanding genetic abnormalities encouraged British biologist Robert Edwards to study the maturation of oocytes. In 1965, he published details of the maturation of human oocytes observed outside the body.[24] In this article he discussed the value of his research for the alleviation of infertility. The main problem in Edwards' research was getting human eggs. Trying to solve this difficulty, he contacted Oldham gynecologist Patrick Steptoe, a pioneer in the use of the technique of laparoscopy in Britain. Their collaboration allowed Edwards to obtain oocytes from some of Steptoe's infertility patients. By 1969, when they published a paper in *Nature*, they had worked with 73 eggs.[25] Of these eggs, the authors claimed that sperm penetrated 18 and some had gone through the preliminary stages of post-fertilization cell division. They estimated that most of the 18 eggs had been fertilized. By the next year, they could get eggs to divide to the blastocyst stage.[26]

In the early 1970s, Edwards and Steptoe in Britain, Pierre Soupart in the United States, and Carl Wood in Australia were attempting human *in*

---

22. J. Rock and M. F. Menking, "In Vitro Fertilization and Cleavage of Human Ovarian Eggs," *Science* 100 (1944): 105-107.

23. M. C. Chang, "Fertilization of Rabbit Ova in Vitro," *Nature* 184 (1959): 466-467.

24. R. G. Edwards, "Maturation in Vitro of Human Ovarian Oocytes," *The Lancet* 2 (1965): 926-929.

25. R. G. Edwards, B. D. Bavister and P. C. Steptoe, "Early Stages of Fertilization in Vitro of Human Oocytes Matured in Vitro," *Nature* 221 (1969): 632-635.

26. R. G. Edwards, P. C. Steptoe, and J. M. Purdy, "Fertilization and Cleavage in Vitro of Preovulatory Human Oocytes," *Nature* 227 (1970): 1307-1310.

vitro fertilization and embryo transfer. In 1973, Carl Wood's group reported an unsuccessful implantation of an 8-cell embryo. The pregnancy ended after nine days when the abdominal wound ruptured.[27] On July 15, 1978, the work of Edwards and Steptoe led to the birth of the first human after conception *in vitro*. They did not use ovulation induction in this first attempt. The child was conceived after a laparoscopic aspiration of an oocyte during a natural cycle.[28] Quickly after this first attempt, specialists realized that they could obtain more mature and fertilizable oocytes and improve pregnancy rates if they used pharmacologic ovarian stimulation.[29] At the same time, the widespread use of gonadotropins for ovarian induction led to a more refined and convenient means for monitoring ovarian response. In 1979, Ylostalo *et al.* reported on the use of ultrasound during stimulation of the ovaries.[30]

Over the next few months, medical practice throughout the Western world incorporated IVF technology. Australian researchers developed more effective egg retrieval methods and adjusted fertility drug doses and combinations to increase the number of oocytes produced.

Another significant event of the 1980s was the development in Australia of embryo cryopreservation and thawing techniques.[31] A. Trounson and L. Mohr described the first human pregnancy resulting from the transfer of a cryopreserved embryo. This pregnancy ended in spontaneous abortion.[32] In 1984 G.H. Zeilmaker and his colleges

---

27. D. de Kretzer *et al.*, "Transfer of a Human Zygote," *The Lancet* 2 (1973): 728-9.

28. P. C. Steptoe and R. G. Edwards, "Birth after the Reimplantation of a Human Embryo," *Lancet* 2 (1978): 366; and R. G. Edwards and P. C. Steptoe, *A Matter of Life: The Story of a Medical Breakthrough* (London: Hutchinson, 1980).

29. See, for example, A. Lopata *et al.*, "In Vitro Fertilization of Preovulatory oocytes and Embryo Transfer in Infertile Patients Treated with Clomiphene and Human Chorionic Gonadotropin," *Fertility and Sterility* 30 (1978): 27-35.

30. P. Ylostalo, L. Ronnberg, and P. Jouppila, "Measurement of the Ovarian Follicle by Ultrasound in Ovulation Induction," *Fertility and Sterility* 31 (1979): 651-655.

31. See, for example, A. Trounson and J. Shaw, "The Cryopreservation of Human Eggs and Embryos," in *Reproductive Medicine and Surgery*, eds., Wallach and Zacur, pp. 860-868; X. J. Wang, "The Contribution of Embryo Cryopreservation to In Vitro Fertilization/Gamete Intrafallopian Transfer: 8 Years Experience," *Human Reproduction* 9:1 (1994): 103-109.

32. A. Trounson and L. Mohr, "Human Pregnancy Following Cryopreservation, Thawing and Transfer of an Eight-Cell Embryo," *Nature* 305 (1983): 707-709.

reported the first live birth using this procedure.[33] The use of frozen embryos has the advantage of allowing professionals to retrieve multiple eggs from the woman's ovaries at once. Specialists can transfer one or several embryos to the uterus and freeze the others for future use if necessary.

Also, in the first part of 1980s, the advent of IVF allowed for the donation of oocytes and embryos from one woman to another.[34] The first successful pregnancy with an embryo donation occurred in 1984.[35] The original indication for using these techniques was premature ovarian failure or severe genetic abnormalities in the woman. At present, the most common receivers of donated eggs are post-menopausal women.[36]

Modification and variations from the basic IVF procedure occurred rapidly. R.H. Ash and colleagues introduced the technique of gamete intrafallopian transfer (GIFT) in 1985.[37] In this method, the specialists transfer both eggs and sperm to the woman's tubes. This technique is now widely used in patients with at least one normal fallopian tube.[38]

---

33. G. H. Zeilmaker *et al.*, "Two Pregnancies following Transfer of Intact Frozen-Thawed Embryos," *Fertility and Sterility* 42 (1984): 293-296.

34. L. J. Trounson and C. Wood, "The Use of Donor Eggs and Embryos in the Management of Human Infertility," *Aus N Z J Obstet Gynaecol* 24 (1984): 265-270.

35. P. Lutjen *et al.*, "The Establishment and Maintenance of Pregnancy Using In Vitro Fertilization and Embryo Donation in a Patient with Primary Ovarian Failure," *Nature* 307 (1984): 174-175.

36. See, for example, M. V. Sauer *et al.*, "A Preliminary Report on Oocyte Donation Extending Reproductive Potential to Women over 40," *New England Journal of Medicine* 323 (1990): 1157-1160; S. Antinory *et al.*, Oocyte Donation in Menopausal Women," *Human Reproduction* 8 (1993): 1487-1490; O. K. Davis and Z. Rosenwaks, "Donor Oocytes in Assisted Reproduction," in *Reproductive Medicine and Surgery*, eds., Wallach and Zacur, pp. 869-877; N. L. Dean and R. G. Edwards, "Oocyte Donation -- Implications for Fertility Treatment in the Nineties," *Current Opinion in Obstetrics and Gynecology* 6:2 (1994): 160-165; and Y. Yaron *et al.*, "In Vitro Fertilization and Oocyte Donation in Women 45 Years of Age and Older," *Fertility and Sterility* 63:1 (1995): 71-6.

37. R. H. Ash *et al.*, "Gamete Intrafallopian Transfer (GIFT): a New Treatment for Infertility," *International Journal of Fertility* 30 (1985): 41-45. See also, R. H. Ash *et al.*, "Preliminary Experiences with Gamete Intrafallopian Transfer (GIFT)," *Fertility and Sterility* 45 (1986): 366-371.

38. See, for example, T. Tanbo, P. O. Dale, and T. Abyholm, "Assisted Fertilization in Infertile Women with Patent Fallopian Tubes: a Comparison on In Vitro Fertilization, Gamete Intrafallopian Transfer, and Tubal Embryo Stage Transfer," *Human Reproduction* 5 (1992): 266; H. Abramovici *et al.*, "Gamete

Many practitioners consider GIFT a better method because the embryos generated in this manner develop in the protective environment of the tube and enter the uterus in a natural way.

Another variation of IVF is zygote intrafallopian transfer (ZIFT). In this procedure, about 18 hours after fertilization, specialists place the zygote in the fallopian tube.[39] Devroey and his associates reported the first successful pregnancy from ZIFT in 1986.[40] Shortly thereafter, Yovich *et al.* reported a new technique of zygote intrafallopian transfer called pronuclear stage tubal transfer (PROST).[41] In this method, doctors deliver the pronuclear oocyte (i.e., newly fertilized ova) into the fallopian tube.

As IVF and related techniques developed in the laboratory and in clinical practice, organizations representing the medical, legal, and religious communities, as well as many women's groups, began to examine the ethical, social, and legal issues surrounding these technological procedures. In the next section, I give a brief account of the different strategies that these groups have used in their evaluations of IVF and related procedures.

1.5. In Vitro Fertilization: Ethical and Social Issues

In the early 1970s, before it become clear that *in vitro* fertilization was a clinical reality, discussion of its ethical aspects had already arisen. Arguments supporting and rejecting the use of IVF, both in research and

---

Intrafallopian Transfer. An Overview," *Journal of Reproductive Medicine* 38 (1993): 698-702; T. Abyholm and T. Tanbo, "GIFT, ZIFT, and Related Techniques: *Current Opinion in Obstetrics and Gynecology* 5 (1993): 615-22; and J. P. Balmaceda, A. Manzur, and R. H. Asch, "Gamete Intrafallopian Transfer," in *Reproductive Medicine and Surgery*, eds., Wallach and Zacur, pp. 806-817.

39. See, for example, M. R. Fluker, C. G. Zouves, and M. W. Bebbington, "A Prospective Randomized Comparison of Zygote Intrafallopian Transfer (ZIFT) Versus Standard IVF-ET for the Treatment of Nontubal Factor Infertility," *Fertility and Sterility* 60 (1993): 515-519; and American College of Obstetricians and Gynecologists (ACOG) Committee Opinion, "Zygote Intrafallopian Transfer," *International Journal of Gynaecology and Obstetrics* 42 (1993): 74.

40. P. Devroey *et al.*, "Pregnancy after Laparoscopic Zygote Intrafallopian Transfer in a Patient with Sperm Antibodies [Letter]," *Lancet* 1 (1986): 329.

41. J. L. Yovich *et al.*, "Pregnancies following Pronuclear Stage Tubal Transfer," *Fertility and Sterility* 48 (1987): 851-857.

in infertility treatments, were present in some of the main scientific and social journals.[42] During these initial years, the issues more intensely and frequently discussed centered on the morality of embryo experimentation, the risk of clinical IVF research to the potential child, and the possible long-term consequences of the investigation on traditional social values such as the family.

Arguments against the use of IVF were varied. Some of those who opposed IVF suggested the need for further animal experimentation before clinical application of IVF. They also considered that IVF research was unethical because the procedures could carry serious risks of damage for the child generated; researchers could not exclude the possibility of irreparable damage to the child-to-be. Therefore, they argued, scientists could not morally proceed with the research. Opponents of IVF also called attention to its possible long-term consequences on society. They warned of the donation of ova or embryo to third parties, the gestation of embryos in surrogate mothers, sex determination, genetic engineering, cloning, an extracorporeal gestation. The introduction of third-party gametes into the relationship of an established couple could, according to these critics, pose potential psychological dilemmas for the infertile partners. It also could present problems surrounding disclosure or nondisclosure to children of the circumstances surrounding their conception. The donation of embryos and the use of surrogate mothers raised objections of possible custody conflicts and exploitation of poor women coerced to act as surrogates or as donors.[43]

---

42. For a review of published articles on the subject see L. Walter, "Human In Vitro Fertilization: A Review of the Ethical Literature," *Hastings Center Report* 9:4 (1979): 23-43.

43. See, for example, P. Ramsey, *Fabricated Man: The Ethics of Genetic Control* (New Haven: Yale University Press, 1970); L.R. Kass, " Babies by Means of In Vitro Fertilization: Unethical Experiments on the Unborn? *New England Journal of Medicine* 285 (1971): 1174-1179; J. D. Watson, "Potential Consequences of Experimentation with Human Eggs," in Committee on Science and Astronautics, *International Science Policy* (Washington, D.C.: U.S. Government Printing Office, 1971); P. Ramsey, "Shall We `Reproduce'? I. The Medical Ethics of In Vitro Fertilization," *JAMA* 220 (1972): 1346-1350; P. Ramsey, "Shall We `Reproduce'? II. Rejoinders and Future Forecast," *JAMA* 220 (1972): 1480-1485; L. R. Kass, "Making Babies--the New Biology and the `Old' Morality," *The Public Interest* 26 (1972): 18-56; M. Lappé, "Risk-Taking for the Unborn," *Hastings Center Report* 2:1 (1972); I. Jakobovits, *Jewish Medical Ethics*, 2nd ed. (New York: Block, 1975); B. Häring,

Supporters of IVF presented also diverse arguments. They claimed that IVF could solve the problems of many infertile couples. They asserted that the desire to have children was a basic human instinct and that denying it could lead to psychological and social difficulties.[44] They argued that IVF, at least in the context of heterosexual marriage, did not pose any moral problems. Those who favored IVF also maintained that the benefits of the application of the technique --giving infertile couples their own children-- outweighed the costs --treating woman and experimenting on embryos. Similarly, they argued that the quantity and quality of animal research on IVF were sufficient to justify its clinical application and that animal research suggested that the procedure did not seem to increase the risks of malformations. As for the long-term impacts of the application of IVF, some champions of the technique welcomed the liberation from coital-gestational reproduction, the possibility of reducing the incidence of genetic defects, and the clinical value of sex determination. They also maintained that ova donation was ethical when a woman lacked ovaries or feared passing on a genetic disease. Some supporters of IVF considered embryo donation also ethically acceptable in cases of cardiac disorders, partial paralysis, or repeated miscarriages.[45]

After the July 1978 birth of the first child as a result of *in vitro* fertilization, religious organizations, especially the Roman Catholic Church and "pro-life" groups who resisted human intervention in the reproductive process, fervently opposed IVF and related technologies. According to these organizations, such procedures severed the natural link between sexual intercourse and procreation and, therefore, they were

---

Ethics of Manipulation: Issues in Medicine, Behavior Control, and Genetics (New York: Seabury Press, 1975).

44. See R. G. Edwards and D. J. Sharpe, "Social Values and Research in Human Embryology," *Nature* 231:5298 (1971), pp. 87-91 (hereafter cited as: Edwards and Sharpe, Social Values).

45. See, for example, Edwards and Sharpe, Social Values; F. P. Grad, "New Beginnings in Life. A Lawyer Response," in *The New Genetic and the Future of Man*, ed. M. P. Hamilton, (Grand Rapids, Michigan: William B. Berdmans, 1972), pp.64-77; G. F. B. Schumacher *et al.*, "In Vitro Fertilization of Human Ova and Blastocyst Transfer. An Invitational Symposium," *The Journal of Reproductive Medicine* 11:5 (1973): 192-200; R. G. Edwards, "Fertilization of Human Eggs In Vitro: Morals, Ethics, and the Law," *The Quarterly Review of Biology* 49:1 (1974), pp. 3-26; J. Fletcher, *The Ethics of Genetic Control: Ending Reproductive Roulette* (Garden City, New York: Anchor Press/Doubleday, 1974).

impermissible. They objected to IVF because it separated the origin of life from acts of interpersonal love. These critics were also concerned with the destruction and manipulation of embryos because they considered human life, from the moment of fertilization, worthy of moral respect and legal protection. Thus, manipulating fertilized eggs or discarding them was tantamount to abortion or human experimentation without consent.[46]

In the mid-1980s, however, feminists' criticisms began to dominate opposition to IVF and related procedures. They claimed that the medical profession was taking control of female reproductive capacities. In 1984, in Groningen, the Netherlands, the participants at the Second International Interdisciplinary Congress on Women created FINNRET (Feminist International Network on New Reproductive Technologies). In 1985, women from FINNRET, concerned with the advances of new reproductive procedures, organized the Women's Emergency Conference on New Reproductive Technologies in Vallinge, Sweden. At this conference, the participants emphasized the link between genetic engineering and reproductive techniques. They pointed out the different meaning and implications of the new reproductive technologies for women in developed and developing nations, as well as their harmful effects for women now and in the future. Participants also stressed the need for feminist resistance strategies to all these procedures. Thus, they changed the name from FINNRET to FINRRAGE (Feminist International Network of Resistance to Reproductive and Genetic Engineering). The main goals of this group were: (1) to assess the impacts of reproductive and genetic engineering on women's life; (2) to raise public awareness about these techniques; (3) to monitor international development in the areas of reproductive and genetic medicine; (4) to examine the relationship between science, technology, and social relations underlying these new procedures; and (5) to create a

---

46. See, for example, L. Kass, "'Making Babies' Revisited," *Public Interest* 54 (1979): 32-60; B. Mitchell, *Morality: Religious and Secular* (Oxford: Clarendon Press, 1980); W. May, "Begotten Not Made: Reflections on Laboratory Production of Human Life," in *Perspectives in Bioethics*, ed. F. Lescoe (New Britain, CT: Mariel, 1983), pp. 29-60; J. Santamaria and N. Tonty-Filippini, eds., *Proceedings of the 1984 Conference in Bioethics* (Melbourne: St. Vincent's Bioethic Center, 1984); Congregation for the Doctrine of the Faith, *Instruction on the Respect for Human Life in Its Origin and on the Dignity of Procreation* (Boston: St. Paul Editions, 1987).

global movement of feminist resistance toward reproductive and genetic research areas.[47]

These feminist groups called attention, for the first time, to the effects of IVF and related techniques on women. Religious, pro-life, and professional critics were worried about the consequences of the new procedures for the embryo, the fetus, or for traditional values such as the family. Feminists, on the other hand, stressed the risks that these techniques posed to women's health as well as their impacts on women's status in society. According to these feminist groups, the new procedures were not designed to give women more choices, as their supporters claimed. On the contrary, they were based on the capitalist and patriarchal ideology of abusing, exploiting, and failing to respect women. These groups also warned about the dismemberment of women's bodies, the commercialization of motherhood, and the eugenic and racist biases that the new technologies were promoting. They emphasized the necessity for identifying the causes of infertility and for working toward prevention as a more rational and appropriate way of solving women's reproductive problems.[48]

After the initial feminist criticisms of new assisted-conception techniques, other feminist scholars started to question their view of the procedures as an unmitigated attack on women. FINRRAGE certainly opened the doors for a debate centered on women. However, other

---

47. See, for example, P. Spallone and D.L. Steinberg, eds., *Made to Order: The Myth of Reproductive and Genetic Progress* (Oxford, England: Pergamon Press, 1987); L. Woll, "The Effect of Feminist Opposition to Reproductive Technology: A Case Study in Victoria, Australia," *Issues in Reproductive and Genetic Engineering* 5 (1992): 21-38.

48. See, for example, R. Arditti, R. Klein, and S. Minden, eds., *Test-Tube Women. What Future for Motherhood?* (London: Pandora Press, 1984); G. Corea, *The Mother Machine: Reproductive Technologies from Artificial Insemination to Artificial Wombs* (New York: Harper and Row, 1985); G. Corea *et al.*, eds., *Man-Made Women. How Reproductive Technologies Affect Women* (London: Hutchinson, 1985); J. A. Scutt, ed., *Baby Machine. Reproductive Technology and the Commercialization of Motherhood* (Melbourne: McCulloch Publishing, 1988); P. Spallone, *Beyond Conception: The New Politics of Reproduction* (London: Macmillan, 1989); R. Klein, *The Exploitation of a Desire. Women's Experiences with In Vitro Fertilization* (Geelong: Deaking University Press, 1989); R. Rowland, *Living Laboratories. Women and the New Reproductive Technologies* (Bloomington: Indiana University Press, 1992); R. Klein, *The Ultimate Colonization: Reproductive and Genetic Engineering* (Dublin: Attic Press, 1992).

feminist groups suggested a different position. They argued that assisted-conception techniques could be used to women's advantage. Although they recognized that no technology is neutral, they rejected the social and technological determinism that some FINRRAGE members seemed to suggest. These feminist critics also acknowledged that the social policy surrounding the assisted-conception techniques harmed women's interests. However, they opposed the image of women as brainwashed individuals immersed in a world of constructed needs and unable to decide by themselves. They promoted a widespread public discussion and eventual political and legislative action to improve women's reproductive autonomy instead of a complete rejection of the new procedures.[49]

During the first years of the debate, doctors and other proponents of IVF and related technologies based their justification for using and developing them mainly on the importance of the desire and need to have children.[50] They expressed their claims about needs in terms of social, biological, or psychological necessity.[51] The media emphasized the grief and despair of childless women and the medical profession took the role

---

49. See, for example, M. Stanworth, ed., *Reproductive Technologies. Gender, Motherhood, and Medicine* (Minneapolis: University of Minnesota Press, 1987); C. Overall, *Ethics and Human Reproduction. A Feminist Analysis* (Boston: Allen & Unwin, 1987); C. Overall, ed., *The Future of Human Reproduction* (Toronto: The Women's Press, 1989); L. Birke, S. Himmelweith, and G. Vines (eds.), *Tomorrow's Child* (London: Virago Press, 1990). See, also, S. Behuniak-Long, "Radical Conceptions: Reproductive Technologies and Feminist Theories," *Women and Politics* 10:3 (1990), pp. 39-64; and N.N. Chokr, "Feminist Perspectives on Reproductive Technologies: The Politics of Motherhood," *Technology in Society* 14 (1992): 317-333.

50. See, for example, P. Singer, and D. Wells, *The Reproduction Revolution* (Oxford: Oxford University Press, 1984) (hereafter cited as Singer and Wells, *Revolution*; and M. Warnock, A *Question of Life: The Warnock Report on Human Fertilization and Embryology* (Oxford: Blackwell, 1985)(hereafter cited as Warnock Report).

51. See, for example, Edwards and Sharpe, Social Values; and J.M. Bardwick, *In Transition* (New York: Holt, Rinehart & Winston, 1979); P. Singer, and D. Wells, *Revolution*; M. French, *Beyond Power: Women, Men, and Morals* (London: Jonathan Cape, 1985); Warnock *Report*; R. Edwards, *Life Before Birth. Reflections on the Embryo Debate* (New York: Basic Books, 1989) (hereafter cited as R. Edwards, *Life*).

of helping them.[52] Thus, IVF appeared as a cure to a growing disease: infertility.[53]

Infertility, however, seemed to increase with the development of the IVF industry and assisted-conception techniques. Although in the early 1980s IVF appeared as a miraculous technique, research on it was still expensive and controversial. Religious and political opposition obstructed unrestrained experimentation with these new technologies, and investigation could only be funded privately, particularly in the United States. Lack of federal funding did not, however, discourage the pharmaceutical and medical industry from investing considerable amounts of money on the research. As a consequence, hundreds of specialized IVF clinics opened across America, Europe, and Australia.[54] Thus, while professionals glorified IVF, they also expanded the medical indications for which it was applicable. Although the initial use was to bypass infertility due to blocked or missing fallopian tubes, IVF treatment expanded to include cases of unexplained infertility, sperm antibodies, and male-related conditions.[55] Expanding the definition of infertile people

---

52. See, for example, M. Kaplan, "Coping with Infertility," *Boston Magazine* 72 (1980): 114-121; M. Olmstead, "Health: Infertility," *Working Woman* 6 (1981): 34-37; *Newsweek* 100 (1982): 102-110; J.H. Bellina, "The Stress of Infertility," *Cosmopolitan* 198 (1985): 156-160; S. Lewis, "The Emotional Impact of Infertility," *Glamour* 83 (1985): 244-248; H. Brotman, "In Search of a Baby; Fertility Specialists Armed with Sophisticated Equipment and Techniques Offer New Hope and Help for the 2.5 Million Couples Who Need Their Services," *Consumer Digest* 24 (1985): 47-50; and. Lord , "Desperately Seeking Baby: Ten Million Americans Are Struggling to have Children," *U.S. News & World Report* 103 (1987): 58-63.

53. See, for example, J. Michaels, "One out of Every Six Couples Want a Baby and Cannot Have One," *Redbook* 154 (1980): 39-45; R.H. Blank, "The Infertility Epidemic; Childlessness Plagues American Couples," *The Futurist* 19 (1985): 17; D. R. Zimmerman, "The Shocking Reason More and More Young Women Can't Have Babies," *Good Housekeeping* 200 (1985): 26-30; and A. Quindlen, "Baby Craving: Facing Widespread Infertility, a Generation Presses the Limits of Medicine and Morality,: *Life* 10 (1986): 23-26.

54. See, for example, Subcommittee on Regulation, Business Opportunities, and Energy, *Consumer Protection Issues Involving In Vitro Fertilization Clinics* (Washington, D.C.: U.S. Government Printing Office, 1989) (hereafter cited as SRB, *Consumer*, 1989); J. Gunning and V. English, *Human In Vitro Fertilization* (Aldershot: Dartmouth, 1993) (hereafter cited as Gunning and English, *Human)*; Brownlee, Baby Chase; and J. Palca, Word.

55. See, for example, Dawson, K., "Ethical Aspects of IVF and Human Embryo Research," in *Handbook of In Vitro Fertilization*, eds., Trounson and Gardner, pp.

who could use IVF and related techniques to introduce these new applications guaranteed the required outlet for what started to be a regular service.

Soon, the media started more and more to frame the need to have a child in terms of moral and legal justifications. The arguments about needs changed into claims about rights, introducing assumptions of legal entitlement.[56] Advocates of IVF and related techniques argued that women have a right to procreate, as stated in various international declarations on human rights.[57] The increased availability of the new reproductive techniques appeared to provide the condition for asserting the right to use them. Thus, some supporters of these technologies, such as Dr. Edwards, argued that the pain caused by involuntary childlessness demanded treatment for the couples and funding of IVF research. People, they claimed, have a right to benefit from research whenever possible as well as a right to receive treatment if they ask for it.[58]

At the end of the 1980s, IVF became a common product whose need seemed to be beyond dispute. Longer waiting lists for IVF and related technologies resulted as a combination of several factors: (1) more clinics and more physicians interested in infertility services, (2) an alleged rise in reproductive problems in the population, and (3) an extension of the medical conditions for which specialists use these techniques. Patient needs now became consumer demands.[59]

The increase in the demand did not, however, result in the perfection of IVF and related techniques and, in the 1990s, consumer protection appears necessary in the U.S. and elsewhere. The procedures have low

---

287-302; P. M. Lewis, "Patient Selection and Management," in *A Textbook of in Vitro Fertilization and Assisted Reproductive Technology*, eds., Brinsden and Rainsbury, pp. 27-37; Chen and Wallach, Five; and Damewood, In Vitro;

56. See chapter five for problems on how IVF assessors have treat issues of individual rights vs. common good.

57. R. Edwards and P. Steptoe, *Matter*, C. Wood and A. Westmore, *Test Tube Conceptions* (Melbourne: Hill of Content, 1983); J. Robertson, "Procreative Liberty and the Control of Conception, Pregnancy and Childbirth," *Virginia Law Review* 69 (1983): 405-462; and J. Robertson, "Embryos, Family, and Procreative Liberty: The Legal Structures of the New Reproduction," *Southern California Law Review* 59 (1986): 942-1041.

58. See, R. Edwards, *Life*, p. 31.

59. See J. Van Dyck, *Manufacturing Babies & Public Consent Debating the New Reproductive Technologies* (New York: New York University Press, 1995).

success rates (8%-10%) and are highly expensive (approximately $60,000 per attempt). Infertile couples find it extremely difficult to obtain clear, unbiased information about the performance of IVF clinics. Thus, although IVF clinics usually publish up to 20% success rates, some of them have not produced even one live child. Besides, because of the scarcity of government regulation of the IVF industry, doctors can hold themselves out as fertility specialists. Additionally, because of the painful and distressing experience of infertility, people suffering from reproductive problems appear especially vulnerable to exploitation. Thus, they need protection from dangerous or misapplied medical technologies.[60]

In countries other than the U.S., however, especially in those with a publicly funded health care system, the discussion is less centered in terms of consumer protection. In nations like Canada, Australia, or the United Kingdom, the debate about new reproductive technologies raises questions concerning what services should be considered medically necessary, who should have access, and who should pay the costs.

1.6. In Vitro Fertilization: Public Policy

The legal problems that IVF and related technologies pose are manifold and, in most cases, without precedent. Such problems include licensing to IVF clinics and embryo experimentation laboratories, ownership of embryos, or legal status of children born through IVF. The public mechanisms used to cope with such issues vary widely across different countries and states.[61] In this section, I give a brief account of the regulatory approaches to human *in vitro* fertilization and associated techniques that a number of nations around the world have taken. These regulatory efforts represent an amalgam of legislation, reports of commissions, administrative regulations, and court opinions. First, I shall

---

60. See Subcommittee on Regulation, Business Opportunities, *Consumer Protection Issues Involving In Vitro Fertilization Clinics* (Washington, D.C.: U.S. Government Printing Office, 1988; (hereafter cited as SRB, *Consumer*, 1988); SRB, *Consumer*, 1989; and SHE, *Infertility*.

61. See, for example, R.H. Blank, *Regulating Reproduction* (New York: Columbia University Press, 1990) (hereafter cited as R.H. Blank, *Regulating*); and Gunning and English, *Human*.

review the current attempts to manage these procedures in the United States. Next, I shall describe the activities of Australia and Canada, as well as those of Western European countries where there exists some kind of regulation or legislation.

1.6.1 UNITED STATES OF AMERICA

In dealing with the regulatory efforts in the United States, I shall address three main topics: (A) constitutional issues related to the use and possible regulation of IVF and related techniques; (B) federal law applicable to these procedures; and (C) state law.

Constitutional Issues

The extent to which States can ban or regulate assisted-conception technologies depends on the degree to which the U.S. Constitution protects procreation. Thus, if the right to procreate appears in the Constitution as fundamental, then only compelling state interest, i.e., if the state can show great harm to others, can justify a regulation that imposes a burden on such a basic right.[62]

The U.S. Constitution seems to imply the right to procreate, that is, to bear or beget a child. It is implicitly grounded in both individual liberty[63] and the integrity of the family unit,[64] and is regarded as a "fundamental right,"[65] one that is essential to the notion of liberty and justice.[66] The U.S. Supreme Court, however, has not explicitly considered whether there is a positive right to procreate, i.e., whether every individual has a right actually to bear or beget children and therefore has a claim on the community for necessary assistance in such an undertaking. It has, nevertheless, considered a wide range of related issues, including the duty of a state not to interfere with the individual's procreative capacities by forcible sterilization,[67] the individual's rights to

62. *Carey v. Population Servs.* Int'l, 431 U.S. 678, 686 (1977).
63. *Skinner v. Oklahoma*, 316 U.S. 535 (1942).
64. *Meyer v. Nebraska*, 262 U.S. 390 (1923).
65. *Griswold v. Connecticut*, 381 U.S. 479 (1965).
66. *Eisenstadt v. Baird*, 405 U.S. 438 (1972).
67. *Skinner v. Oklahoma*, 316 U.S. 535 (1942).

prevent conception or to terminate a pregnancy,[68] and the right of individuals to rear children in nontraditional family groups.[69]

In 1942, in a decision striking down Oklahoma's compulsory sterilization law, the Supreme Court held that individuals have a right to be free from unwarranted governmental interference with procreative capabilities. The *Skinner* decision is notable for its explicit mention of "a right to have offspring." Although the Constitution does not specifically mention procreation, the court characterized it as a basic human right.

This attitude has continued to pervade subsequent court decisions. In 1965, the court held in *Griswold v. Connecticut* that married couples have a right to be free of state interference in their decision to obtain and use contraceptives. The court based its resolution on the concept of marital privacy, another area constitutionally protected from government interference.[70] But the court did not restrict to married couples the right to determine whether or not to procreate. In 1972, the court explicitly extended this principle to individuals when it held that an individual's right to obtain and use contraceptives was also protected by the right to privacy.[71] Similarly, in cases upholding the right to terminate a pregnancy, the court has made clear that such a right extends to both married and unmarried. In the *Roe v. Wade* decision the court extended the individual's right to privacy in procreation to a woman's right to decide whether to terminate a pregnancy during the first trimester.[72]

The series of cases from *Skinner* to *Roe* suggest that a right to reproduce exists. But such a right appears to be negative, i.e., as a right to be free from unwarranted governmental interference. In the absence of a state interest that overrides people's reproductive choices, they have the right to procreate as long as their action does not harm a constitutionally defined person. The right, therefore, denotes freedom of choice relating to whether or when to procreate.

The development of assisted-conception techniques, however, raises the possibility of extending the right to reproduce. According to scholars such as J. Robertson, if the right to procreate raises a strong presumption against unwarranted interference, this presumption applies equally to a

---

68. *Roe v. Wade*, 410 U.S. 113 (1973).

69. *Moore v. City of East Cleveland*, 431 U.S. 494 (1977).

70. *Griswold v. Connecticut*, 381 U.S. 479 (1965).

71. *Eisenstadt v. Baird*, 405 U.S. 438 (1972).

72. *Roe v. Wade*, 410 U.S. 113 (1973).

person's freedom to choose IVF and related procedures. Infertile couples, these critics assert, have as meritorious a need to reproduce as fertile individuals. They should have the same right to have and rear offspring through the assistance of medical technology as fertile people have through sexual intercourse. Therefore, supporters of procreative rights maintain, state power that goes beyond facilitating regulation of these procedures should be extremely limited. A governmental ban on assisted-conception techniques, they say, would be unconstitutional.[73]

Some scholars have argued that common-good considerations may sometimes override procreative liberties. According to these theorists, those who give priority to individual rights to procreate often underestimate the impacts of assisted-conception techniques such as IVF on women, children, or the community as a whole.[74]

Federal Law

The IVF industry has operated virtually without federal rules since 1975, when the Department of Health, Education, and Welfare (now the Department of Health and Human Services) published regulations relating to IVF. The provision that had the most important effect in keeping the federal government out of funding, and thereby reviewing, IVF research was:

> No application or proposal involving human *in vitro* fertilization may be funded by the Department or any component thereof until the application or proposal has been reviewed by the Ethical Advisory Board and the Board has rendered advice as to its acceptability from an ethical standpoint.[75]

In 1977, the Department of Health, Education, and Welfare (HEW) received an application for research involving human *in vitro* fertilization. The investigation proposed to remove eggs from women undergoing

---

73. See, for example, J. A. Robertson, "Liberty and Assisted Reproduction," *Trial* 30:8 (1994), pp.49-53; and J. A. Robertson, *Children of Choice* (Princeton: Princeton University Press, 1994), ch. 2.

74. See, for example, C. Overall, *Ethics and Human Reproduction. A Feminist Analysis* (Boston: Allen & Unwin, 1987), ch. 8; and M. Ryan, "The Argument for Unlimited Procreative Liberty: A Feminist Critique," *Hastings Center Report*, vol. 6 (July/August 1990).

75. 45CFR46.204(D)

surgery; researchers would then fertilize the eggs, with subsequent microbiopsy of the embryos. In September of that year, the Department appointed an Ethics Advisory Board (EAB) to evaluate the proposal from a scientific point of view. At a meeting in May 1978 the Board agreed to review it. In July of that year, however, R. Edwards and P. Steptoe announced the birth of the first baby following IVF. In September of 1978 Secretary Califano asked the Board to consider the social, scientific, ethical, and legal issues surrounding IVF research. Over the next nine months the Board held a series of eleven public hearings in which private individuals, professional societies, and public-interest groups presented their opinions. Scholars and experts in the field of reproductive science, ethics, theology, law, and the social sciences prepared reports for the Board. In addition, it received over two thousand pieces of correspondence that were copied and distributed to each of the members. In May of 1979 the EAB issued its report.[76]

The Board concluded that research involving human *in vitro* fertilization without embryo transfer was acceptable from an ethical point of view provided that: (1) the research complied with all appropriate provisions of the regulations governing research with human subjects[77]; (2) it was designed primarily to establish the safety and efficacy of embryo transfer and to obtain further scientific knowledge about it; (3) human gametes came only from persons who have been informed of the nature and purpose of the research; (4) professionals did not sustain embryos beyond 14 days after fertilization; and (5) specialists advised all interested parties if the evidence showed that the procedure entailed risks of abnormal offspring. The Board also concluded that if the IVF research involved embryo transfer, doctors should use only gametes obtained from lawfully married couples, and they should replace the fertilized ova back into the woman from whom they came. The EAB found it ethically acceptable for the Department to support or conduct research involving *in vitro* fertilization and embryo transfer, although it decided not to

76. See Ethics Advisory Board, Department of Health, and Welfare, *Report and Conclusions: HEW Support of Research Involving In Vitro Fertilization and Embryo Transfer* (Washington, D.C.: U.S. Government Printing Office, 1979) (hereafter cited as EAB, *Report*). See, also, Gunning and English, *Human*, p. 18-22.

77. 45CFR46

address the question of the level of funding, if any, for such investigations.[78]

In spite of the report that the EAB offered, the Department of Health, Education, and Welfare took no action. The Board's charter expired in 1980 and both the Reagan and Bush Administrations failed to renew it. Without a Board, however, government could not fund IVF research, so there was a *de facto* moratorium. The effect of this moratorium was to eliminate the most direct line of authority by which the federal government could influence the development of embryo research and infertility treatments to avoid unacceptable practices or inappropriate uses.[79]

At the end of the 1980s, the growing demand for IVF services, the proliferation of IVF clinics, together with a low success rate (10%) and evidence of misinformation to the public, started to concern the federal government. Since 1987 separate House subcommittees have held preliminary hearings on diverse aspects of IVF and related techniques. In 1987, the Select Committee on Children, Youth, and Families took testimony from various experts on assisted-conception procedures in order to explore the medical, ethical, and legal issues with a particular focus on their implications for families and children.[80] In 1988 and 1989, hearings of the Subcommittee on Regulation and Business Opportunities addressed specifically the question of regulation. Chair Ron Wyden's concern with the lack of adequate federal regulation motivated these hearings. Participants at the hearings recognized the value of the standards that professional organizations such as the American Fertility Society and the American College of Obstetricians and Gynecologists promulgate. However, they realized that voluntary guidelines were not strong enough to produce a self-enforcement of IVF services. They emphasized the possibility of exploitation of infertile couples because of

78. See EAB, *Report*, p.106-108.

79. See, for example, OTA, *Infertility*; T. Powledge, "Reproductive Technologies and the Bottom Line," in *Embryos, Ethics, and Women's Rights* eds., E.H. Baruch, A. F. D'Adamo, and J. Seager (New York: Harrington Park Press, 1988), pp. 203-209; Gunning and English, *Human*, pp. 18-22; and Palca, Word, p. 5; and U.S. Congress, Office of Technology Assessment, *Biomedical Ethics in U.S. Public Policy* (Washington, D.C.: U.S. Government Printing Office, 1993), p. 11.

80. Select Committee on Children, Youth, and Families, *Alternative Reproductive Technologies: Implications for Children and Families* (Washington, D.C.: U.S. Government Printing Office, 1987).

an inappropriate use of credentials, misuse of new technologies, and misleading advertising. The Chair highlighted two areas of essential action: dissemination of information to consumers and regulation of the IVF industry.[81]

In 1992, U.S. Congressmen Ron Wyden and Norman Lent introduced a Bill to offer the public comparable information concerning effectiveness of infertility services. The bill proposes to assure the quality of such services by providing for the certification of embryo laboratories. The *Fertility Clinic Success Rate and Certification Act of 1992* requires all infertility clinics that perform IVF services to communicate annually their pregnancy success rates to the Secretary of Health and Human Services. It demands as well the identity of each embryo laboratory working in association with the clinic. The Bill also directs the Secretary to develop a model program for State certification of embryo laboratory accreditation programs. In addition, it demands that the Secretary publishes and disseminates data concerning pregnancy success rates and other related information. The Senate passed the Bill on October 8, 1992. President Bush signed it on October 24 of the same year. The effective date of the Act is two years after the date of enactment of the legislation.[82] The Centers for Disease Control and Prevention continue developing the actual mechanisms for the implementation of the *Fertility Clinic Success Rate and Certification Act of 1992*.[83]

The situation regarding funding IVF also changed with the National Institute of Health Revitalization Act of 1993. The Act states that the provisions of the 45CFR204(d) shall not have any legal effect. D. Alexander, director of the National Institute of Child Health and Human Development, decided shortly after the Congress passed the Act to revisit the guidelines about funding of embryo experimentation that the Ethics Advisory Board had laid down in 1979. He also created an *ad hoc* advisory panel to address the topic of IVF and embryo research. In October 1994, an advisory panel to the director of the National Institute

---

81. See SRB, *Consumer*, 1988; and SRB, *Consumer*, 1989.

82. See, SHE, *Fertility*; U.S. Congress, House of Representatives, *Fertility Clinic Success Rate and Certification Act of 1992* (Washington, D.C.: U.S. Government Printing Office, 1992), Report 102-624; and U.S. Senate, *Fertility Clinic Success Rate and Certification Act of 1992* (Washington, D.C.: U.S. Government Printing Office, 1992), Report 102-452.

83. See CDC, *Success Rate*.

of Health proposed lifting the ban on funding for a variety of researches on human *in vitro* fertilization. The committee established a new set of guidelines to govern such research, limiting it to embryos 14 days or less from first cell division. It banned payments to donors of gametocytes or embryos. Also, the committee approved the production, use, and disposal of embryos by *in vitro* fertilization solely for research purposes. However, due to a constitutional challenge these measures are not in effect yet.[84]

State Law

Statutory regulation of IVF does not differ much from the initial action of the federal government. Only five states, Illinois, Kentucky, Louisiana, New Mexico, and Pennsylvania, specifically mention IVF. Of them, only Pennsylvania and Louisiana explicitly address therapeutic IVF. In 1982, Pennsylvania enacted the first IVF statute. The statute does not impede the practice of IVF, but the record-keeping and reporting provisions allow the state to monitor it. The penalty prescribed for failure to submit the report is a fine. In 1986, Louisiana passed a law that restricted practice of IVF to those facilities that comply with the guidelines of the American Fertility Society or the American College of Obstetricians and Gynecologists. The law forbids the purposeful creation of an in vitro embryo solely for research or sale. It also specifically bans the sale of human ova. Besides it grants the status of "juridical person" to the fertilized ovum, until the time of implantation into the uterus. As a juridical person, the fertilized ovum may sue or be sued and may not be considered property. The law considers the gamete donors its parents, and if they are unidentifiable, the facility is its guardian. Also, no person, including the donors, may intentionally destroy an in vitro embryo that appears capable of normal development. Kentucky, on the other hand, addresses IVF in its adoption statutes only to say that nothing in the law banning adoption in certain instances prohibits IVF. Illinois' fetal research statute forbids the sale of and experimentation on feta but allows performance of IVF. Finally, a New Mexico statute defines "regulated research" to include that involving IVF, but not IVF performed to treat

---

84. See J. Palca, Word; C. Marwick, "NIH Panel Finds Embryo Research Justifiable, Recommends Support," *JAMA* 272:17 (1994), pp. 1311-1312; and E.D. Pellegrino, "Ethics," *JAMA* 273:21 (1995), pp. 1674-1676.

infertility. Thus this law effect on IVF as a clinical therapy does not appear to be significant.[85]

Because of the lack of legislative directives on IVF and related techniques, courts have to deal with the resolution of the legal issues such as parental relationships, possible ownership of gametes and embryos, conflicts of custody, and so on. One of the consequences of this lack of legislative directives is that, because the 50 states are free to come to their own conclusions concerning IVF rules, there may be significant State-to-State variations in the law.[86]

Regulatory efforts in other countries seem to be a little more successful than in the United States. In what follows we offer a summary of some of such endeavors in different nations around the world.

1.6.2. AUSTRALIA

Australian specialists undertook pioneering work on *in vitro* fertilization and there has been considerable activity in the country surrounding IVF and related technologies.[87] Such activity includes federal and state reports, state legislation, and professional self-regulation.

The Federal Government has published three reports on new reproductive techniques. In 1985, the Senate Standing Committee on Constitutional and Legal Affairs issued a report on *IVF and the Status of Children*. The study analyzes the lack of uniformity in previous legislation establishing the status of IVF children as well as evaluates the significance of this lack. In the same year, the Family Law Council of the Attorney-General's Office published an account examining assisted-conception procedures in Australia. The report, *Creating Children: A Uniform Approach to the Law and Practice of Reproductive Technology in Australia,* states the need for uniform law throughout the country regarding the status of children born using donor gametes. Also in 1985,

---

85. See, for example, OTA, *Infertility*, pp. 250-252; R.H. Blank, *Regulating*, pp. 119-122; and M.F. Milich, "In Vitro Fertilization and Embryo Transfer: Medical Technology + Social Values = Legislative Solutions," *Journal of Family Law* 30:4 (1991-1992): 875-901; and Gunning and English, *Human*, pp. 166-169.

86. See previous note.

87. See, for example, OTA, *Infertility*, pp. 329-332; R.H. Blank, *Regulating*, pp. 154-159; Gunning and English, *Human*, pp.144-147

Senator Harradine (Tasmania) introduced a *Human Embryo Protection from Experimentation Bill,* which led to a Senate Select Committee enquiry. The Committee solicited written submissions from a wide range of individuals and organizations with expertise and interest on the topic. In 1986, the Committee offered its final report, *Human Embryo Experimentation in Australia*.

Australia also has a national regulatory system concerned with ethical aspects of clinical IVF and embryo research. The National Health and Medical Research Council (NHMRC), a body charged with advising federal, state, and territory governments, directs this system. In 1985 the NHMRC offered its guidelines in the *NH&MRC Statement on Human Embryo Experimentation and Supplementary Notes*. In these rules, the NHMRC agreed that IVF is a justifiable means of treating infertility. Nevertheless, they also stated that much research still needed to be done. In the same year, the NHMRC undertook a review of the 12 IVF centers then operating in Australia to establish whether they were observing their instructions. The Council did not find significant departures from the directives.

At the state level, Victoria passed legislation to regulate IVF in November 1984. In 1988 and 1991, respectively, South Australia and Western Australia followed Victoria. Although other states have held committees of enquiry into IVF and related techniques, they have made no attempts to legislate solely on IVF. However, all the states as well as the Australian Capital Territory and the Northern Territory have legislation on the legal status of children conceived by assisted-conception procedures.

The Victorian Government set up a Committee of Enquiry after the birth in 1979 of the first Australian child conceived *in vitro* in this state. The Committee under the direction of Professor Louis Waller produced three reports over two years. It presented its *Interim Report* in September 1982, the *Report on Donor Gametes in Vitro Fertilization* in August 1983, and the *Report on the Disposition of Embryos Produced by in Vitro Fertilization* in August 1984. Following the recommendations on these reports, the Victorian Government enacted the *Infertility (Medical Procedures) Act* in 1984. The Bill set out the provisions to regulate IVF practice. It allows the fertilization of ova outside a woman's body only for implantation and only for married couples. Counseling is mandatory and couples may only enter an IVF program provided that twelve months have passed since another medical practitioner indicated IVF to be appropriate treatment. It limits the practice of IVF to approved hospitals and provides for record keeping and confidentiality. The Act also provides for the establishment of a Standing Review and Advisory Committee to consider requests for approval of embryo research proposals and to advise the minister on infertility.

The Standing Review Committee was established in 1985 also under the chair of Professor Louis Waller. On its recommendations, the Parliament amended the Act in 1987 to include GIFT and related techniques. The amended Act also permits research on the process of fertilization before syngamy (the final stage of fertilization in which chromosomes from the female and the male gametes come together to form the zygote) on eggs not destined for replacement.

The next state to enact legislation in the Commonwealth of Australia was South Australia. The *Reproductive Technology Act 1988* establishes a Council on Reproductive Technology of 11 members to monitor IVF practice and research, and to advise the minister on issues of reproductive procedures. It lays down provisions for licensing of IVF centers and research by the South Australian Health Commission. The Bill allows the use of artificial fertilization procedures only for married couples (defined as couples having cohabited for at least five years). It also prohibits both any researches that may be detrimental to the embryos and their cryopreservation beyond ten years.

Western Australia passed legislation on assisted-conception techniques in 1991.[88] The *Human Reproductive Technology Act 1991* received assent on October 8, 1991. The Bill establishes the Western Australia Reproductive Technology Council to compile a code of practice to govern the use of artificial fertilization procedures. The Council also advises the Minister of Health on matters related to reproductive techniques and promotes informed public debate and education on such issues. The aim of the legislation is to make these techniques available to infertile couples and those whose children are likely to be affected by genetic disease. The Act does not allow the fertilization of eggs, other than for implantation in the body of a woman.

---

88. See, for example, OTA, *Infertility*, pp. 329-332; R.H. Blank, *Regulating*, pp. 154-159; P. Singer *et al.*, eds., *Embryo Experimentation* (Cambridge: Cambridge University press, 1990), pp. 227-230, (hereafter cited as Singer, *Embryo*); Gunning and English, *Human*, pp. 144-147

It also prescribes that gamete providers have rights of control over their own gametes and must consent to keeping or using them to create an embryo.

1.6.3. CANADA

There has been considerable debate in Canada over a number of years concerning new reproductive technologies.[89] However, as yet there is no Canadian legislation pertaining to IVF. In 1985, the Ontario Law Reform Commission published the *Report of Human Artificial Reproduction and Related Matters*. It considers IVF acceptable in principle, although only for clinical reasons. The Commission also accepts the donation of eggs and embryos.

In 1992, the Law Reform Commission of Canada issued the working paper *Medically Assisted Procreation*. This report addresses the medical and legal aspects of reproductive techniques, makes 22 recommendations, and sets out the proposed contents of a Medically Assisted Procreation Act. This Act provides that access to IVF and related treatments be denied to no one unless costs or scarcity of resources require a selection. It allows the donation of gametes and embryos with the written consent of the donors and with such consent withdrawable in writing as well. It prohibits all commercialization of gametes and embryos and forbids doctors to select specific qualities unless for the purpose of preventing the transmission of serious genetic disease. The proposed Bill also restricts clinical and storage services to certified clinics and gamete or embryo banks. It requires every clinic to provide counseling for patients and to keep written annual reports and file them with a central registry. Finally, it recommends the creation of a national regulatory agency to certify clinics and banks and ensure compliance with the Act.

In 1989, the Canadian Government established a Royal Commission on New Reproductive Technology under the chair of Dr. Patricia Baird. The Commission's mandate was to inquire into and report on current and potential scientific and medical developments related to assisted-conception procedures. Its aim was to pay special attention to the social, legal, health, research, ethical, and economic implications of these

---

89. See, for example, OTA, *Infertility*, pp. 332-335; R.H. Blank, *Regulating*, pp. 154-159; and Gunning and English, *Human*, pp. 149-150

techniques and to recommend what policies to apply. Since then, the Commission has undertaken a broad exercise of public consultation and research. The Commission was due to report back to the government in 1992 but there was a delay until 1993 partly because of the Commission's decision to extend the consultation period and partly because of hostilities among its members. In 1993, *Proceed with Care*, the final report of the Royal Commission on New Reproductive Technologies, appeared.[90] They recommended legislation to prohibit, with criminal sanctions, several aspects of the new techniques such as using embryos in research related to cloning and animal/human hybrids. They proposed banning the use of eggs from female fetuses for implantation; the sale of eggs, sperm, embryos, or fetal tissues; and advertising for, paying for, and acting as an intermediary for surrogacy arrangements. The Commission also recommended that the federal government establish a regulatory and licensing body -- the National Reproductive Technologies Commission (NRTC). The NRTC would have five areas of regulatory responsibility in which the provision of services would be subject to compulsory licensing through five subcommittees established for that purpose. These areas are (1) sperm collection, storage, and distribution, and the provision of assisted-insemination services; (2) assisted-conception services; (3) prenatal diagnosis; (4) research involving embryo experimentation; and (5) the provision of human fetal tissue for research. The Commission recommended also the establishment of a sixth subcommittee with primary responsibility in the field of infertility prevention.

1.6.4. AUSTRIA

In 1986, the Ministry of Sciences and Research published the report *The Fundamental Aspects of Genetics and Reproductive Biology*, intended to be a foundation for legislation.[91] The 1992 Act No. 275 regulating medically assisted procreation requires reproductive procedures to be used as a treatment for infertility only when other treatments have failed and only for couples married or in stable relationships. According to the

---

90. Royal Commission on New Reproductive Technologies, *Proceed with Care* (Ottawa, Canada: Canada Communications Group, 1993).

91. See, for example, OTA, *Infertility*, pp. 346; and Gunning and English, *Human*, pp. 147-148

Act, specialists may fertilize eggs solely for implantation. If there are surplus embryos (spare embryos that doctors do not transfer into the woman's body initially) practitioners may freeze them for up to one year, but only the couple from whom the gametes originated can use the embryos. The Act does not allow embryo research. Counseling is mandatory and, where a couple is unmarried or donor sperm needed, the Act requires counseling before a notary regarding the legal consequences of consent. The Bill also contains conscience clauses allowing physicians and nursing staff to refuse to participate in medically-assisted conceptions.

1.6.5. DENMARK

In 1987, the Danish Parliament established a National Ethics Council.[92] The Council's main objective was to make recommendations on embryo research, pre-implantation diagnosis, and gene therapy. It started its job in January 1988 and, after two years of public consultation, it presented its report: *Protection of Human Gametes, Fertilized Ova, Embryos and Fetuses*. The Parliament debated the report and drafted and subsequently amended a Bill in 1991. The resulting Act, No. 503 of June 24, 1992, came into force in October of the same year, with a statutory obligation for its review and revision, if necessary. The Act accepts that IVF is an acceptable infertility treatment and that research is necessary for its further development. It prohibits the donation of fertilized eggs as well as cloning, mixing genetically different pre-embryos and embryos, and the production of hybrids between human beings and animals. The Act also requires that investigators report any biomedical research project involving human individuals, human gametes intended for use in fertilization, pre-embryos and embryos to regional ethics committees. Denmark has seven of these committees that oversee and approve research within the terms of the Act.

1.6.6. FRANCE

The Conseil d' Etat, a permanent advisory body to the government in matters of legislation in 1988, started its analysis of reproductive

---

92. See, for example, Singer, *Embryo*, pp. 230-231; Gunning and English, *Human*, pp. 150-152.

technologies.[93] It recommended, in its proposal for legislation, a ban on the creation of embryos for research, although it allowed strictly limited research on embryo surpluses. It also recommended that both progenitors give written consent to the research. In addition, the Conseil d'Etat recommended that the National Ethics Committee, whose aim is to provide advice on ethical issues raised by medical and biological investigations and to promote public debate, should authorize the investigation. The Conseil d'Etat advised against any research leading to the modification of the genome, complete gestation *in vitro*, or cloning.

A draft bill materialized in April 1986 following the recommendations of the Conseil d' Etat. Consequently, the Prime Minister commissioned a further report to offer an in- depth study of recent developments in biology, assisted reproduction, and genetics. The goal was to provide an international overview of the technology available together with a summary of the bioethical approaches adopted. The report *Aux frontières de la vie: une éthique biomedicale à la française*, appeared in 1991. It prohibits the creation of embryos specifically for research and the implantation of embryos subject to experimentation. It also recommends that specialists obtain consent from donors before using surplus embryos for investigation. In March 1992, the French Parliament introduced a bill to address the clinical provisions of assisted-conception procedures. The bill states the terms for licensing clinics and laboratories and the creation of a National Council for the Biology and Medicine of Reproduction.

1.6.7. GERMANY

In 1984, the Federal Ministry for Research and Technology initiated discussion of the ethical implications of biotechnologies and, in collaboration with the Ministry of Justice, considered restrictions on noncoital reproduction.[94] In 1985, they completed their joint report, known as the Benda report and recommended numerous restrictions on the use of assisted-conception techniques. The report advises the use of IVF only for married couples utilizing their own gametes; although it also says that in exceptional circumstances cohabiting couples could use IVF. The Benda Report recommends that legislation be enacted restricting the

---

93. See, for example, Gunning and English, *Human*, pp. 152-153.

94. See, for example, OTA, *Infertility*, pp. 335-338; and Gunning and English, *Human*, pp. 154.

use of IVF and related techniques to medical establishments that satisfy specific safety requirements. The medical profession, however, opposed the legislation and insisted on the sufficiency of self-regulation.

In 1989, the German Parliament started to discuss legislation on new reproductive technologies and in January 1991 the *Embryo Protection Act* came into force. The Bill is entirely restrictive and prohibits fertilization other than for the purposes of pregnancy; fertilization of more eggs than may be necessary for transfer in one attempt; the transfer of more than three eggs or embryos in one treatment cycle; and fertilization of a human egg for any purpose other than to begin a pregnancy in the woman that produced the ovum. Other punishable offenses include cloning, manipulation of gametes, and the formation of human/animal hybrids. In all cases the practitioner, not those seeking treatment, is punishable under the Act.

1.6.8. GREECE

Currently there is no legislation in Greece regulating assisted conception.[95] However, a law on the modernization of the health system introduced in July 1992 paves the way. Such law makes provisions for a presidential decree that will regulate the conditions under which units for assisted reproduction may be established and may function. The same decree will regulate deontological, ethical, economical, and legal aspects of the technologies. Assisted-fertilization units would be able to function only in the context of specifically organized private or public hospitals or private clinics. The decree also makes a provision for the creation of a national ethics committee.

1.6.9. NETHERLANDS

In 1983, following a request from the Minister of Health, the Health Council established a commission to look into IVF and embryo research.[96] In 1984, the Commission published an interim report suggesting making IVF available only in cases of poor tubal pathology. It

---

95. See, for example, OTA, *Infertility*, pp. 348; and Gunning and English, *Human*, pp. 158-160.

96. See, for example, OTA, *Infertility*, pp. 350-351; and Gunning and English, *Human*, pp. 158-160.

also recommended developing assisted-conception services primarily in academic hospitals. The final report, *Artificial Procreation*, appeared in 1986 and in its conclusions includes donor insemination, oocyte donation, and surrogacy. In its latest report *Heredity: Science and Society*, published in 1989, the Health Council considers the issue of research on human embryos. It draws up some procedural stipulations according to which specialists cannot grow an embryo *in vitro* beyond 14 days of development; they may use IVF only when they cannot obtain by other means the information gained from the experiments with embryos; and they cannot implant embryos used for research. The report also recommends a moratorium on human germ-line gene therapy.

The Ministers of Health and Justice established the regulation of IVF by decree of August 1988. It announced that the Government intended to bring IVF fully under the Hospital Facilities Act and to maintain the licensing system that the Minister of Health had proposed in 1985.

1.6.10. NORWAY

The Norwegian *Act No. 68 of 12 June 1987 Relating to Artificial Procreation* limits IVF to specially approved institutions and only for married couples.[97] It forbids donor gametes, research on fertilized eggs, and the freezing of unfertilized ova. It specifies that clinics may store embryo surpluses for 12 months and that doctors may not treat patients over 37 years of age. The legislation requires institutions carrying out artificial fertilization to report fully to the Ministry of Social Affairs. To engage in prohibited conduct is an offence punishable by a fine or imprisonment.

1.6.11. PORTUGAL

Decree-Law No. 319/86 (September 1986) is rather general.[98] However, it provides that the manipulation, collection, and observation of sperm, and any other procedures required by the techniques of assisted reproduction, should be done under the supervision of a physician in authorized public or private facilities.

---

97. See, for example, Singer, *Embryo*, pp. 232; and Gunning and English, *Human*, pp. 160-162.

98. See, for example, Singer, *Embryo*, pp. 232.

1.6.12. SPAIN

In April 1986, Marcel Palacios, President of a Special Commission set up by the Cortes to study human *in vitro* fertilization and donor insemination, reported on these two issues.[99] The commission presented 155 recommendations for legislative and regulatory action about infertility treatments. Following this report, Spanish Law 35, *Health: Assisted Reproduction Techniques* received the Royal Assent on November 22, 1988. The Law covers artificial insemination, IVF, and GIFT. It lays down general principles for the application of these technologies that emphasize informed consent, full disclosure of risks, patient data collection and confidentiality, fertilization of ova for the sole purpose of procreation, and the minimization of spare embryos. The Law also permits sperm and embryo freezing but prohibits ova freezing until the technique is proven to be safe for thawed eggs. It specifies conditions applicable to gamete donors, persons undergoing the treatment, and the status of resultant children. The law makes these services available to any women, whether married or not. It also decrees that a National Commission for Assisted Reproduction shall regulate infertility services. The Spanish Fertility Society established an IVF register to collect statistics, starting 1989.

1.6.13. SWEDEN

In 1981, the Swedish Government formed a committee to investigate the issues surrounding noncoital reproductive technologies.[100] The Insemination Committee published two reports, one in 1983 concerning artificial insemination and one in 1985 concerning IVF and surrogate motherhood. The recommendations of the 1985 report resulted in Swedish Law No. 711 of June 14, 1988 on *Fertilization outside the Human Body* that came into force on January 1, 1989. The Act sets forth the conditions under which specialists may undertake IVF and embryo research. It makes IVF available only to married couples or couples living together in a permanent relationship. It forbids the use of donated

---

99. See, for example, OTA, *Infertility*, pp. 353-354; Singer (eds.), *Embryo*, pp. 233; and Gunning and English, *Human*, pp. 163-164.

100. See, for example, OTA, *Infertility*, pp. 342-343; P. Singer *et al.*, (eds.), *Embryo*, pp. 233; and Gunning and English, *Human*, pp. 164-165.

gametes or the donation of surplus embryos to another couple. The law also bans surrogate motherhood. It permits freezing of surplus embryos for up to one year and only with the couple's consent. Participants may also consent to the use of spare embryos for research. It also allows research up to two weeks following fertilization but only for investigation related to the improvement of IVF procedures. The law provides that, unless otherwise authorized, the treatment must be done in a general hospital. Fine or prison sentences are the punishment for breaching the law.

1.6.14. SWITZERLAND

Legislation on assisted-reproductive technologies in Switzerland is far from homogenous.[101] Of the Cantons that have introduced it, the French and Italian speaking Cantons allow homologous IVF. Of the German speaking ones, only Aargau has legislated in favor of homologous IVF for married couples. The Cantons of Glarus and Basel allow only the use of artificial insemination by the husband and prohibit the utilization of all other methods of assisted conception.

All of the Cantons that permit *in vitro* fertilization do so strictly in accordance with the guidelines of the Swiss Academy of Medical Sciences. These guidelines state that only married couples or unmarried couples living in a stable relationship should be able to use assisted-reproduction techniques. They allow gamete donation provided it is free of charge and the number of offspring from one donor is limited to ten. In addition, the guidelines recommend that only accredited physicians may practice IVF, and only in medical establishments equipped with the relevant technical facilities to perform this procedure.

1.6.15. UNITED KINGDOM

In July 1982, the British Parliament established the Warnock Committee to examine the social, ethical, and legal implications of human-assisted reproduction. On June 25, 1984, under the direction of its chair, Mary Warnock, the Committee presented the *Report of the Committee of Inquiry into Human Fertilization and Embryology* (Warnock Report).

---

101. See, for example, Singer *Embryo*, pp. 234; Gunning and English, *Human*, pp. 165-166.

The report outlined a number of recommendations bearing on IVF. It advised creation of a statutory licensing authority (SLA) to regulate both research and infertility services. The Committee recommended that artificial insemination by donor (AID), IVF, egg and embryo donation, the clinical use of frozen embryos, and research on human embryos be permitted only under licensing. It discouraged the use of embryo donation by lavage and the utilization of frozen embryos until more research was available. The Warnock report also allowed the sale or purchase of human gametes and embryos only under license from the SLA.

Besides the recommendations regarding licensing and control, the Warnock committee proposed principles of provision designed to protect the public interest. Some of these were: sufficient public funding for the collection of adequate statistics on infertility and infertility services; establishment of a working group at the national level to draw up detailed guidance for the organization of services; anonymity of donors; availability of counseling to all couples and third parties; accessibility of the child upon reaching age 18 to information about the donor's ethnic origin and genetic health; and informed consent of all parties. The Warnock committee also recommended some legal changes. Many of them deal with clarifying the legitimacy of children born through AID and IVF. Other recommendations involve specifics of succession and inheritance of people who decide to use assisted-conception techniques. It also recommended (although with an expression of dissent) the introduction of legislation to render criminal the creation of surrogacy agencies and all surrogacy arrangements.

The proposal to create an interim licensing body received approval from the Medical Research Council (MRC) and the Royal College of Obstetricians and Gynecologists (RCOG). In March 1985, they established the Voluntary Licensing Authority (VLA) whose main tasks would be to register and approve centers undertaking IVF and to consider and approve proposals for research.

The Government's first response to the Warnock Report was to introduce legislation --the *Surrogacy Arrangement Act 1985*-- banning commercial surrogacy. The Act dealt only with surrogacy. Regarding other issues surrounding infertility treatment, the government decided to look for further consultation. Thus, in 1986, the Department of Health and Social Security released a Consultation Paper, *Legislation on Human*

Infertility Services and Embryo Research. The document encouraged further discussion of the need for a statutory licensing authority for infertility treatment as well as for the storage, disposal, and research on human embryos. It also called attention to the need to counsel infertile couples and to the legal status of children resulting from techniques that use donated gametes.

The consultation period ended in June 1987, and in November 1987 the government issued a White Paper, *Human Fertilization and Embryology: A Framework for Legislation*, that should be the basis for future regulation. This proposal generally followed the recommendations of the Warnock Committee. The White Paper proposed a statutory licensing authority and set out its function, composition, and licensing powers.

The *Human Fertilization and Embryology Act 1990* brings about the eventual implementation of the Warnock proposals. The Act establishes the Human Fertilization and Embryology Authority (HFEA) to cover the statutory licensing of clinical IVF, the donation and storage of gametes, and embryo research. The HFEA took up its full responsibilities on August 1, 1991. The Act allows licensed research, but it does not permit keeping or using embryos beyond 14 days after fertilization. It also prohibits cloning and the placing of a human embryo in any other animal. It amends the *Surrogacy Arrangements Act 1985* to make all surrogacy arrangements unenforceable and extends it to include the use of all new reproductive technologies.[102]

1.7. Summary and Conclusion

In order to understand the ethical, social, legal, and political impacts that IVF and related technologies may have in our society, governments often have turned to institutional committees for assessment of these procedures. This chapter has offered a brief description of IVF as a clinical procedure to solve infertility problems. I also have presented an account of the scientific development of this technique and a discussion of some ethical and social issues raised by the implementation and use of IVF. Finally, the chapter summarized some legislative and regulatory

---

102. See, for example, OTA, *Infertility*, pp. 343-345; Singer, *Embryo*, pp. 234; and Gunning and English, *Human*.

practices related to IVF in the United States, Austria, Canada, and several western European countries.

In the forthcoming chapters, I analyze some of the inadequacies in the evaluation of IVF and argue that these deficiencies may have wrongly guided public policy about these procedures. I show some of the epistemological and ethical problems that assessors of IVF and related technologies have disregarded, and analyze some of the consequences that follow from such omissions.

Implementing and unrestrictedly using new technologies without adequately assessing them is not uncommon. Use of x-rays, DES, Thalidomide, and nuclear power are examples of this practice. They are also examples of procedures that have had disastrous consequences in our society. At the least, we should try to learn from our mistakes and attempt not to commit them again. This analysis is one step in our learning and our avoiding ethical and policy errors.

CHAPTER 4

ETHICS AND UNCERTAINTY: IN VITRO FERTILIZATION AND RISKS TO WOMEN'S HEALTH

1.1. Introduction

From the 1940s to the early 1970s, doctors gave diethylstilbertrol (DES) to millions of pregnant women in order to prevent spontaneous abortions. Physicians did not have, however, adequate information about the potential consequences of the use of this hormone. Years later, researchers have found that two to four million daughters of women who took DES are suffering cancer of the vagina and cervix at a rate higher than that of the female population their own age. These women also have experienced increased rates of infertility, spontaneous abortions, and ectopic (outside the uterus) pregnancies. Moreover, more than thirty years after they used the drug, women who took DES are suffering from 40% to 50% higher rates of breast cancer than other women of their age.[1] Similarly, the case of the Dalkon Shield, an intrauterine-contraceptive device marketed for several years, exemplifies problems in dealing with medical technologies. Between 1970 and 1974, doctors inserted the Dalkon Shield into more than two million U.S. women and a

---

1. See, for example, C. Orenberg, *DES: The Complete Story* (New York: St. Martin's Press, 1981); R. Klein and R. Rowland, "Women As Test Sites for Fertility Drugs: Clomiphene Citrate and Hormonal Cocktails," *Reproductive and Genetic Engineering. Journal of International Feminist Analysis*, 1: 3 (1988): 251-73; Harriet Simand, "138-188: Fifty Years of DES --Fifty Years Too Many," in *The Future of Human Reproduction*, ed., C. Overall (Toronto: The Women's Press, 1989), pp.95-104; R. Rowland, *Living Laboratories* (Bloomington: Indiana University Press, 1992), p. 50 (hereafter cited as Rowland, *Laboratories*); and S. M. Fisher and R. J. Apfel, "Diethylstilbestrol and Infertility: The Past, the Present, and the Relevance for the Future," in *Technology and Infertility*, eds., M. M. Seibel, A. A. Kiessling, J. Bernstein, and S. R. Levin (New York: Springer-Verlag, 1993), pp. 413-423.

total of four million worldwide. After women used it extensively, researchers compiled evidence indicating that severe hemorrhaging, infected miscarriages, ectopic pregnancies, infertility, mutilated reproductive organs, and death could result from its utilization.

Advances in the biomedical sciences are helping to cure diseases, give children to the infertile, improve the quality of life, and increase longevity. But some of these biomedical developments are causing death and injury, creating alarm among the public, and confronting society with new ethical dilemmas about reproductive rights, parenthood, medicalization of reproduction, exploitation, consent, and equity in the distribution of scarce medical resources.

1.2. Overview

This chapter argues that because of inadequate technology assessments, policymakers have made decisions, in relation to *in vitro* fertilization (IVF), that may not be in the best interests of the public. I defend the thesis that assessments of IVF are inadequate because, in neglecting epistemological and ethical problems such as choosing criteria for decisions under uncertainty, assessors in the four reports may encourage public policies that overlook the possibility of jeopardizing women's health. In section three, I offer an account of the empirical data on IVF risks. I also show that assessors have neglected or underestimated the evidence on IVF hazards. In section four, I argue that evaluators also have undervalued the insufficiency of investigations on IVF risks. In section five, I present a brief explanation of two main decision rules used in situations of uncertainty: expected-utility maximization and maximin. I discuss these decision rules as presented by two of their more important supporters, J. Harsanyi as a proponent of expected-utility maximization and J. Rawls as an advocate of maximin. In section six, I argue that evaluators have erred in their analysis because, underestimating both the existing scientific evidence and the insufficiency of data on IVF safety and effectiveness, they have condoned the use of IVF. Thus, they have implicitly sanctioned criteria such as expected-utility maximization for deciding in a situation of scientific uncertainty. They also have recommended minimizing false positives over false negatives. As a consequence, many women may be exposed to needless risks. In section seven, I argue that, in sanctioning the expansion of IVF to an increasing number of reproductive conditions, assessors also have implicitly sanctioned questionable criteria (expected utility) for making decisions under uncertainty. As a result, women's health may be endangered. Finally, section eight tries to answer some possible objections against my arguments. One of those objections is that because evaluators have recommended obtaining written informed consent from women as a way to overcome problems with risks, their recommendations undercut my worries about medical threats to women. A second criticism refers to the fact that interfering with women's ability to choose a risky technology, such as IVF, might be paternalistic.

1.3. Underestimating In Vitro Fertilization Risks

The issue of whether women undergoing IVF may be exposed to serious risks is important not only for scientific reasons but also for basic ethical, political, and social ones. Assessment of the existing scientific evidence on IVF safety and efficiency is therefore crucial in order to provide input to policymakers. In this section, I offer a brief account of the data on IVF risks. Next, I argue that evaluators on the four commissions seem to have underestimated the importance of IVF hazards.

1.3.1. IN VITRO FERTILIZATION RISKS

According to empirical evidence, risks to women undergoing IVF treatment vary from simple nausea to death. For example, the hormones that doctors use to stimulate the ovaries are associated with numerous side effects. Some studies assert that ovulation induction may be a risk factor for certain types of hormone-dependent cancers. Researchers have associated excessive estrogen secretion with ovarian and breast carcinoma, and gonadotropin secretion with ovarian cancer.[2] A

2. See, for example, J. Jarrel, J. Seidel, and P. Bigelow, "Adverse Health Effects of Drugs Used for Ovulation Induction," in *New Reproductive Technologies and the*

substantial body of experimental, clinical, and epidemiological evidence indicates that hormones play a major role in the development of several human cancers.[3] The ability of hormones to stimulate cell division in certain organs, such as the breast, endometrium, and the ovary, may lead (following repeated cell divisions) to the accumulation of random genetic errors that ultimately produce cancer. Hormone-related cancers account for more than 30% of all newly diagnosed female cancer in the United States.[4] Hence any technique (like IVF)--that relies on massive doses of hormones--may be quite dangerous.

The ovarian hyperstimulation syndrome (OHSS) is another possible iatrogenic (caused by medical treatment) consequence of ovulation induction. Women with the severe form of OHSS may suffer renal impairment, liver dysfunction, thromboembolic phenomena, shock, and even death. The incidence of moderate and severe OHSS in IVF treatment ranges from 3% to 4%, quite a high risk, taking into account that IVF is a selective procedure that treats a non-life-threatening condition. This syndrome is extremely rare following natural conception.[5]

Health Care System. The Case for Evidence-Based Medicine, Royal Commission on New Reproductive Technologies (Ottawa, Canada: Canada Communications Group, 1993), pp. 453-549 (hereafter cited as J. Jarrel *et al*, Adverse Health Effects); and P. Stephenson, "Ovulation Induction During Treatment of Infertility: An Assessment of the Risks," in *Tough Choices*, eds., P. Stephenson and M. G. Wagner (Philadelphia: Temple University Press, 1993), pp.97-121 (hereafter cited as Stephenson, Ovulation); A. Brzezinski, *et al*, "Ovarian Stimulation and Breast Cancer: Is There a Link?" *Gynecol Oncol* 52: 3 (1994): 292-5; and R. E. Bristow and B. Y. Karlan, "The Risk of Ovarian Cancer after Treatment for Infertility," *Curr Opin Obstet Gynecol*, 8: 1 (1996): 32-7.

3. See, for example, S. Fishel and P. Jackson, "Follicle Stimulation for High-Tech Pregnancies: Are We Playing It Safe?" *British Medical Journal* 299 (1989): 309-11; Stephenson, Ovulation, pp. 105-107.

4. See, for example, H. P. Schneider and M. Birkhauser, "Does Hormone Replacement Therapy, Modify Risks of Gynecological Cancers?" *Int. J. Fertil. Menopausal Stud.*, 40, suppl. 1 (1995): 40-53; T. J. Key, "Hormones and Cancer in Humans," *Mutat Res*, 333: 1-2 (1995): 59-67; F. Berrino *et al.*, "Serum Sex Hormone Levels after Menopause and Subsequent Breast Cancer," *J Natl Cancer Inst*, 88: 5 (1996): 291-6.

5. See, for example, B. Rizk, "Ovarian Hyperstimulation Syndrome," in *A Textbook of in Vitro Fertilization and Assisted Reproductive Technology*, eds., P. R.

The procedures that doctors normally use to obtain women's eggs, i.e., laparoscopy and ultrasound-guide oocyte retrieval also pose risks to them. Although there are no accurate statistical data about hazards associated with these two procedures, risks related to these technologies include postoperative infections, punctures of an internal organ, hemorrhages, ovarian trauma, and intrapelvic adhesions.[6] Furthermore, intrapelvic adhesions can exacerbate preexisting infertility or cause it in healthy women who undergo IVF treatments when their male partners have reproductive difficulties.[7] Thus IVF could increase or produce the very problem for which women use it as treatment.

Implantation of embryos or gametes into women's bodies also may be hazardous for them. Some of the possible risks are perforation of organs and ectopic pregnancies. Studies show that 5% to 7% of all IVF pregnancies implant outside the uterus.[8] The hazard in the general

---

Brinsden and P. A. Rainsbury (Park Ridge, NJ: The Parthenon Publishing Group, 1992), pp. 369-383; I. Calderon and D. Healy, "Endocrinology of IVF," in *Handbook of In Vitro Fertilization*, eds., A. Trounson and D.K. Gardner (Boca Raton: CRC Press, 1993), pp. 2-16; M. P. Steinkampf and R. E. Blackwell, "Ovulation Induction," in *Textbook of Reproductive Medicine*, eds., B. R. Carr and R. E. Blackwell (Norwalk, Connecticut: Appleton & Lange, 1993), pp. 469-480; Stephenson, Ovulation; J. G. Schenker and Y. Ezra, "Complication of Assisted Reproductive Techniques," *Fertility and Sterility* 61:3 (1994): 411-422; and J. G. Schenker, "Ovarian Hyperstimulation Syndrome," in *Reproductive Medicine and Surgery*, eds., E. E. Wallach and H. A. Zacur (St. Louis: Mosby, 1994), pp. 649-679.

6. See, for example, Rowland, *Laboratories*, pp. 25-30; L. Koch, "Physiological and Psychosocial Risks of the New Reproductive Technologies," in *Tough Choices*, eds., Stephenson and Wagner, pp. 122-134 (hereafter cited as Koch, Physiological); and P. J. Taylor, P. J. and J. V. Kredentser, "Diagnostic and Therapeutic Laparoscopy and Hysteroscopy and Their Relationship to *in Vitro* Fertilization," in *A Textbook of in Vitro Fertilization and Assisted Reproductive Technology*, eds., Brinsden and Rainsbury, pp. 73-92;

7. See P. R. Brinsden, "Oocyte Recovery and Embryo Transfer Techniques for in Vitro Fertilization," in *A Textbook of in Vitro Fertilization and Assisted Reproductive Technology*, eds., Brinsden and Rainsbury , pp. 139-153 (hereafter cited as Brinsden, Oocyte); Rowland, *Laboratories*, pp. 25-30; Koch, Physiological.

8. Medical Research Institute, Society of Assisted Reproductive Technology, The American Fertility Society, "In Vitro Fertilization/ Embryo Transfer in the United States: 1988 Results from the National IVF-ET Registry," *Fertility and Sterility*, 53:13 (1990).

population, however, is approximately 1%.[9] Ectopic gestations may be life-threatening for the woman and can aggravate infertility.[10] Likewise, multiple gestation occurs in about 25% of IVF pregnancies,[11] while it has an incidence of only 2% in the general population.[12] Multiple-birth pregnancies increase the danger of miscarriages, cesarean sections, early labor, and placental dysfunction.

1.3.2. ASSESSMENTS' TREATMENT OF EVIDENCE ON IN VITRO FERTILIZATION RISKS

In spite of the significance of data on IVF risks, particularly those related to cancer, evaluators on the four commissions seem to have undervalued this information. In the Victorian and the British reports there is no significant discussion about the known and suspected dangers of IVF. They describe the steps of the procedure without referring to possible hazards associated with them. For instance, when addressing ovulation induction (one of the steps in IVF), the Victorian report argues that using fertility drugs to stimulate the ovaries during IVF treatments seems "reasonable since pregnancy often followed the use of fertility drugs in women who were not ovulating."[13] According to this report, having ovarian stimulation therapy is almost routine for treating women who

---

9. See O. K. Davis and Z. Rosenwaks, "Assisted Reproductive Technology," in *Textbook of Reproductive Medicine*, eds., Carr and Blackwell, pp. 571-586. See, also, S. F. Marcus and P. R. Brinsden, "Analysis of the Incidence and Risk Factors Associated with Ectopic Pregnancy Following In Vitro Fertilization and Embryo Transfer," *Human Reproduction* 10:1 (1994): 199-203.

10. See Brinsden, Oocyte; Rowland, *Laboratories*, pp.30-32; Koch, Physiological.

11. See, for example, World Health Organization (WHO), "Recommendations on the Management of Services for *in Vitro* Fertilization from the WHO 1990," British Medical Journal, 305 (July 25, 1992): 251 (hereafter cited as WHO, Recommendations); and Stephenson, Ovulation, p. 100.

12. See, for example, Stephenson, Ovulation, p. 100; and S. Brownlee, "The Baby Chase: Millions of Couples Have Infertility Problems, and Many Try High-Tech Remedies. But Who Minds the Price Clinics They Turn to?," *News and World Report* 117 (Dec. 5, 1994): 84.

13. Committee to Consider the Social, Ethical, and Legal Issues Arising from In Vitro Fertilization, *Interim Report* (Victoria: Victorian Government Printer, 1982), p.9 (hereafter cited as Victorian Report).

undergo IVF.[14] Despite the cancer risks, there is no mention, however, of any possible hazards associated with such a therapy. Similarly, the British report presents the use of ovarian stimulation as "very desirable" because it allows the transfer of more than one embryo to the woman's uterus and thus increases the chances of obtaining a pregnancy.[15] Like the Victorian assessment, here there is no discussion whatsoever about cancer risks associated with ovulation-induction drugs.

The Spanish report, on the other hand, does allude to some of the dangers associated with IVF. Nevertheless, it refers to them as "scarce but existent."[16] After this characterization, there is a succinct reference to some of the risks related to IVF. Assessors say that there are some hazards "derived from overstimulation of the ovaries."[17] They forget to mention, however, what those risks are. Next, they say that there are some dangers "derived from general anesthesia, and generated from the surgical procedures used to obtain the eggs, including infections."[18] There is no reference to what kinds of risks are associated with the techniques employed for the retrieval of oocytes or the type of infections these procedures may cause. Evaluators also refer to the risks "derived from multiple-births pregnancies."[19] They say that these kinds of gestations may "produce triplets, quadruplets, etc., with the related obstetric problems (for the woman and children), pediatrics (for the children), and psychological (for the couple)."[20] Again, evaluators fail to mention specific obstetric or pediatric risks, such as use of caesarian sections, early labor, and low-birthweight children, associated with multiple births.

---

14. Victorian Report, p. 9.

15. M. Warnock, *A Question of Life. The Warnock Report on Human Fertilization and Embryology* (Oxford, UK: Blackwell, 1985), pp. 30, 31 (hereafter cited as Warnock Report).

16. Comisión Especial de Estudio de la Fecundación *in Vitro* y la Inseminación Artificial Humanas [Special Commission for the Study of Human *in Vitro* Fertilization and Artificial Insemination], *Informe* [*Report*] (Madrid: Gabinete de Publicaciones, 1987), p. 107 (hereafter cited as Spanish Commission).

17. Spanish Commission, p. 107.

18. Spanish Commission, p. 107.

19. Spanish Commission, p. 107.

20. Spanish Commission, p. 107.

The United States study offers more information about IVF risks than the other reports. It describes some of the potential hazards associated with IVF and related techniques. The U.S. assessment mentions risks produced by the use of fertility drugs such as hyperstimulation; hazards derived from the utilization of techniques to retrieve oocytes such as blood in urine; and dangers attributable to the embryo-transfer procedure such as trauma to the endometrium.[21] It also refers to the fact that "transferred embryos may implant in the fallopian tube."[22] Furthermore, the report says that "the miscarriage rate for infertility patients is generally higher than that for the normal population.[23]" And that "preterm delivery is more common in pregnancies after IVF than in spontaneous pregnancies.[24]" The U.S. study fails to mention, however, any data about cancer risks. When summarizing the main points of the chapter, the report neglects to mention any information at all about hazards.[25]

1.4. Overlooking the Insufficiency of Data on In Vitro Fertilization Risks

Apart from overlooking or underestimating the known risks associated with IVF and related procedures, the commissions also have undervalued the importance of the deficiencies in the existing research and the significance of the scarcity of good investigations on IVF hazards. Assessors do not seem to have taken into account what experts do not know about IVF and related techniques. For instance, evaluators neglect the fact that there is insufficient evidence concerning the long-term effects, such as ovarian and breast cancer, of ovulation-induction drugs in women. Although they recognize the need for improvement in IVF and related procedures, neither the Victorian, the British, nor the Spanish

---

21. Office of Technology Assessment, *Infertility: Medical and Social Choices* (Washington, D.C.: U.S. Government Printing Office, 1988), p.131 (hereafter cited as OTA, *Infertility*).

22. OTA, *Infertility*, p.131.

23. OTA, *Infertility*, p.131.

24. OTA, *Infertility*, p.131.

25. OTA, *Infertility*, p.132.

reports pay attention to the lack of studies on hazards associated with IVF. Only the United States assessment mentions the scarcity of research on maternal health consequences and anomalies in offspring.[26] It does not concede, however, particular significance to this lack of evidence.

Today, more than ten years after the presentation of the first IVF reports (the British and the Victorian), there is a lack of scientifically valid information obtained from clinical trials and other epidemiological investigations on IVF and related techniques. A study conducted in 1993 by the Canadian Royal Commission on New Reproductive Technologies shows that comprehensive evaluations of costs, benefits, risks, and efficiency are still necessary. Systematic data collection is lacking because no organization or agency gathers information on IVF outcomes. For the same reason, little or no follow-up data exist on what happens to children and women after the pregnancies. Record-keeping practices vary markedly among clinics and practitioners, with some clinics not recording data on whether a particular IVF procedure resulted in a live birth.[27] Also, because physicians still do not report all adverse health effects of the ovulation-induction drugs that doctors use in IVF treatment, identification of short- and long-term risks associated with exposure is difficult. Insufficient information about the type of reproductive problem, hormones used, dosage, and duration of infertility treatment make it arduous to assess the impacts of these drugs. Likewise, because of the long latency period of cancer, early detection of the disease is difficult.[28]

---

26. OTA, *Infertility*, pp. 302-303.

27. See Subcommittee on Regulation and Business Opportunities, *Consumer Protection Issues Involving in Vitro Fertilization Clinics* (Washington, D.C.: U.S. Government Printing Office, 1988) (hereafter cited as SRB, *Consumer*, 1988); Subcommittee on Regulation, Business Opportunities, and Energy, *Consumer Protection Issues Involving in Vitro Fertilization Clinics* (Washington, D.C.: U.S. Government Printing Office, 1989) (hereafter cited as SRB, *Consumer*, 1989); Subcommittee on Health and the Environment, *Fertility Clinic Services* (Washington, D.C.: U.S. Government Printing Office, 1992) (hereafter cited as SHE, *Fertility)*; Royal Commission on New Reproductive Technologies, *Proceed with Care* (Ottawa, Canada: Canada Communications Group, 1993), ch. 20 (hereafter cited as RCNRT, *Care*).

28. See, Royal Commission on New Reproductive Technologies, *New Reproductive Technologies and the Health Care System. The Case for Evidence-Based Medicine* (Ottawa, Canada: Canada Communications Group, 1993) (hereafter

Despite the evidence that shows hormones play an important role in the development of ovarian and breast cancer, the overall picture indicates that researchers have displayed very limited interest in investigating the frequency, mechanisms, or impact of the negative health consequences of ovulation-induction therapy.

Investigators have never properly evaluated hormone and IVF-related drugs prior to their introduction into clinical practice. Their omission is serious because, in the past, disastrous consequences have followed from the use of drugs, pesticides, and technologies prior to their assessment. Apart from the mentioned cases of DES, and the Dalkon Shield, other instances of this dangerous practice--"use first, evaluate later"--are the utilization of the pesticide DDT and the implementation of commercial nuclear fission. For example, after years of using DDT, scientists found that this pesticide had the paradoxical consequence of producing a greater pest problem than the one it is supposed to combat. This is because evolution selects for pesticide-resistant insects. The new, stronger insects are more lethal and require more powerful pesticides, which in turn lead to increasing threats to humans and to even more resistant insects.[29] Commercial nuclear fission constitutes another example of implementation of a technology prior to its assessment. Nuclear power has been used for decades in spite of our ignorance of the dangers of low-level nuclear radiation, our knowledge of the probabilities of core melt, and our inability to give a guarantee that disposing of nuclear waste will not harm future generations.[30]

---

cited as RCNRT, *Evidence-Based Medicine).* See, also, SRB, *Consumer*; 1988; SRB, *Consumer*; 1989; SHE, *Fertility*; F. J. Stanley and S. M. Webb, "The Effectiveness of *in Vitro* Fertilization: An Epidemiological Perspective," in *Tough Choices*, eds., Stephenson and Wagner, pp. 62-72; and WHO, Recommendations, p. 251.

29. See, for example, Rachel Carson, *Silent Spring* (Boston: Houghton and Mifflin, 1962); N. Bernard ed., *The Environmental Crisis* (San Diego, CA: Greenhaven Press, 1991); and K. Shrader-Frechette, "Pesticide Toxicity: An Ethical Perspective," in *Environmental Ethics*, ed. K. Shrader-Frechette (Pacific Grove, CA: The Boxwood Press, 1991).

30. See, for example, K. Shrader-Frechette, *Nuclear Power and Public Policy* (Boston: Dordrecht Reidel, 1980); N. Lenssen, "Nuclear Waste: The Problem That Won't Go Away," in *State of the World*, L .R. Brown, (Washington, DC: Norton, 1992); and K. Shrader-Frechette, *Burying Uncertainty* (Berkeley: University of

Likewise, after thirty years of using ovulation-induction hormones, and despite the strong correlation between hormones and cancer, sound investigations of their adverse consequences still are scarce.[31] In comparison, investigators have extensively reviewed drugs administered for chemotherapy and cardiac arrhythmia for their side effects. According to the Canadian study, however, the potential for adverse health consequences in young women using these ovulation-induction drugs is as great as, or greater than, that associated with the treatment of cancer or heart disease.[32] The differences in the investigations of the health impacts of various medications and techniques force one to wonder whether unfair discrimination against women has played a role in the inadequate attention given to the analysis of ovulation-induction drugs. After all, such discrimination has already occurred in other areas of biomedical science, as research into cardiovascular disease illustrates. Although heart trouble is the leading cause of death of women in the U.S., research has concentrated almost entirely on men. Similarly, an investigation demonstrating the effectiveness of aspirin in preventing migraine headaches involved only male subjects, although women outnumber men migraine sufferers three to one.[33]

Assessment, not only of the available scientific evidence, but also of the insufficiency of data on IVF risks, is essential in order to provide information to guide public policy. Nevertheless, assessors have undervalued both of these areas (available research showing risks and

---

California Press, 1993) (hereafter cited as Shrader-Frechette, *Burying*).

31. C. D'Arcy, N. S. B. Rawson, and L. Edouard, "Infertility Treatment -- Epidemiology, Efficacy, Outcomes, and Direct Costs: A Feasibility Study, Saskatcheman 1978-1990," in *Evidence-Based Medicine*, RCNRT, pp. 765-794 (hereafter cited as C. D'Arcy *et al.*, Infertility).

32. Jarrel *et al*, Adverse Health Effects, p. 520.

33. See, for example, H. Lindemann Nelson and J. Lindemann Nelson, "Justice in the Allocation of Health Care Resources: A Feminist Account," in *Feminist and Bioethics*, ed., S. M. Wolf (New York: Oxford University Press, 1996), pp. 351-370 (hereafter cited as H. and J. Lindemann Nelson, Justice). See also, the special report on women's health research in *Science*, 269 (August 11, 1995): 765-801; and A. C. Mastronianni, R. Faden and D. Federman eds., *Women and Health Research: Ethical and Legal Issues of Including Women in Clinical Studies* (Washington, D.C.: National Academy Press, 1994) (hereafter cited as Mastronianni *et al.*, *Women and Health Research*).

scarcity of important data). By underestimating IVF hazards and the insufficiency of information on IVF dangers, evaluators erroneously have presented this biomedical technology as safe. They seem to have assumed that because the evidence has not proved IVF to be risky, then the procedure is safe. Assessors have, therefore, overlooked a third possibility: that IVF threats are uncertain.

1.5. Ethical Alternatives under Uncertainty

We are forced to make decisions numerous times every day. In some cases they are thoughtful, while others are the result only of habit. Because the consequences of our choices affect our lives, we try to make good decisions. However, the conditions under which we have to choose are not always similar. Thus, we may face situations of certainty, risk, and uncertainty. In a situation of certainty we know that a choice will result in a particular outcome; for example, if we use cars that burn gas then we are certain to have the outcome of producing CO_2. In conditions of risk, we know, with a specific probability, whether a choice will result in a given outcome; for instance, when we roll a fair die, we know the particular probability of obtaining a "five." Finally, situations of uncertainty occur when we have partial or total ignorance about whether a choice will result in a given outcome with an assigned probability; for example, if we use ovulation-induction hormones, we have partial ignorance about whether such a choice will result in the probability that at least 10,000 women undergoing IVF treatment will die because of reproductive cancers in a time period of 30 years following the procedure.[34]

Decisions under conditions of certainty and risk are relatively unproblematic because we can attach specific probabilities to given outcomes. Uncertainty situations, however, present more difficulties for our decisions because we do not know the probabilities of particular outcomes.

---

34. See, for example, K. Shrader-Frechette, *Risk and Rationality* (Berkeley: University of California Press, 1991), pp. 101-2, (hereafter cited as Shrader-Frechette, *Risk*).

Because of limited data, questionable theories, or problems of theoretical underdetermination by the evidence, much scientific research is uncertain.[35] Given this uncertainty in science, most technology-related decisionmaking takes place in situations of limited knowledge. People rarely have complete, accurate knowledge of all the probabilities associated with various outcomes of taking technological risks (e.g., ovulation-induction drugs, laparoscopy, ultrasound), because very risky technologies are often new. In order to make policy resolutions, analysts must judge what decision rule would be better to use in a situation of uncertainty. Decision rules are algorithms that unambiguously select the act or acts that are tautologically termed "optimal according to the rule."[36]

There are two main schools of thought about the decision rule that a rational person ought to use in situations of uncertainty. One--the Bayesian School--proposes expected-utility maximization, where utility is a measure of welfare, usually determined by an individual's subjective preferences for a particular state of affairs or set of consequences. Expected-utility is defined as the weighted sum of all possible consequences of the action, where the weights are given by the probability associated with each outcome.[37] The expected utility of an

---

35. See, for example, N. R. Hanson, *Patterns of Discovery* (Cambridge: Cambridge University Press, 1958); M. Polanyi, *Personal Knowledge* (New York: Harper and Row, 1958) K. R. Popper, *The Logic of Scientific Discovery* (New York: Basic Books, 1959); K. R. Popper, *Conjectures and Refutations* (London: Routledge and Kegan Paul, 1963); C. Hempel, *Philosophy of Natural Sciences* (Englewood Cliffs, N. J.: Prentice-Hall, 1966); T. Kuhn, *The Structure of Scientific Revolutions* (Chicago: University of Chicago Press, 1962, 1970);

36. See, for example, R. D. Luce and H. Raiffa, "Individual Decisionmaking under Uncertainty," in *Decision, Probability, and Utility*, eds., P. Gardenfors and N.E. Sahlin (Cambridge: Cambridge University Press, 1988), pp. 48-79. (Hereafter cited as Luce and Raiffa, Decisionmaking.)

37. See, for example, L. J. Savage, *The Foundations of Statistics* (New York: John Wiley & Sons, 1954); and J. Harsanyi, "Can the Maximin Principle Serve as a Basis for Morality? A Critique of John Rawls's Theory," *American Political Science Review* 69, no.2 (1975) (hereafter cited as Harsanyi, Maximin); J. Harsanyi, "Understanding Rational Behavior," in *Foundational Problems in the Special Sciences*, eds., R. E. Butts and J. Hintikka (Boston: Reidel, 1977) (hereafter cited as Harsanyi, Rational); J. Harsanyi, "Advances in Understanding Rational Behavior," in

action--its probable utility--for a case involving two states of affairs is $u_1p + u_2 (1 - p)$, where u_1 and u_2 are outcome utilities, where p is the probability of the state of affairs S_1 and represents the decisionmaker's own subjective probability estimate, and where (1 - p) is the probability of the state of affairs S_2.[38] (With respect to any decision problem under uncertainty, the set of "states of affairs" or "states of nature" is assumed to form a mutually exclusive and exhaustive listing of those aspects of nature which are relevant to a particular choice and about which the decisionmaker is uncertain.)[39] J. Harsanyi takes the Bayesian position that people ought to use expected-utility maximization as the decision rule in situations of uncertainty.

Proponents of maximin (such as J. Rawls) maintain that, in situations of uncertainty, the appropriate decision rule is to maximize the minimum, that is, avoid the choice with the worst possible consequence.[40] Many of the advocates of maximin take it as equivalent to the difference principle. This maximin rule proposes that one society is better than another if the worst-off members of the former do better than the worst-off in the latter.[41] Following is a discussion of these two main decision rules for choices under uncertainty.

1.5.1. HARSANYI AND THE UTILITARIAN STRATEGY

Harsanyi offers three main reasons for the Bayesian/utilitarian strategy and against the maximin principle. His first reason is that those who do not follow expected-utility maximization make irrational decisions because they ignore probabilities. Harsanyi also claims that not following expected-utility maximization leads to unacceptable moral consequences.

Rational Choice, ed. J. Elster (New York: New York University Press, 1986) (hereafter cited as Harsanyi, Advances); and Shrader-Frechette, *Risk*, ch. 8.

38. See Harsanyi, Rational. See, also, Shrader-Frechette, *Risk*, ch. 9.

39. Luce and Raiffa, Decisionmaking, pp. 40-50.

40. See, J. Rawls, *A Theory of Justice* (Cambridge: Harvard University Press, 1971) (hereafter cited as Rawls, *Justice)*; J. Rawls, "Justice as Fairness," *Journal of Philosophy*, 54 (October, 1957): 653-62; and J. Rawls, "Some Reasons for the Maximin Criterion," *American Economic Review*, 64: 1, papers and proceedings (May 1974): 141-46.

41. See Rawls, *Justice*, pp.75-83.

A third reason used by Harsanyi is that, employing the Bayesian/Utilitarian strategy, with the equiprobability assumption, is proper because it allows one to allocate equal a priori probabilities to everyone's interests. (According to the equiprobability assumption, if there is no evidence showing that one event from a complete set of mutually exclusive events is more likely to occur than another, then individuals should judge the events equally probable.)[42] I treat these reasons in order.

According to Harsanyi, in situations of uncertainty, it is irrational to make one's behavior (as the maximin criterion seems to require) wholly dependent on some highly unlike unfavorable contingencies, regardless of how little probability one is willing to assign to them. For Harsanyi, if one chooses the maximin criterion, one could never cross a street, drive over a bridge, or get married, because all these actions could have unpleasant consequences. Thus, according to Harsanyi, in situations of uncertainty, expected-utility maximization suggests policies more reasonable than the maximin principle because one would not make decisions based on highly unlikely adverse outcomes.[43]

Harsanyi also maintains that following maximin instead of expected-utility maximization in situations of uncertainty would lead to unacceptable moral implications. This is the case because the maximin principle would, Harsanyi says, benefit the worst-off individuals, even when they do not deserve it, and even when doing so will not help society.[44]

Harsanyi uses the following examples to show that the maximin principle often has unacceptable moral implications. In the first example, there are two critically ill patients with pneumonia. Their only chance of survival is to be treated by an antibiotic, but there is only enough to treat one patient. One of the patients is a healthy person, apart from the attack of pneumonia. The other is a terminal cancer victim, but the antibiotic could prolong his life for several months. Harsanyi maintains that, maximin supporters would give the antibiotic to the cancer patient because he is the less fortunate of the two patients, whereas Bayesian

---

42. See Harsanyi, Maximin; and Shrader-Frechette, *Risk*, pp.104-16.

43. See Harsanyi, Maximin, pp. 39-40; and Shrader-Frechette, *Risk*, pp.104-08.

44. See Harsanyi, Maximin, pp.40-43.

utilitarians would make the opposite suggestion.[45] In a second example, there are two individuals, a mathematician and a severely retarded person. The question is whether to use their society' s surplus money to provide education for the mathematician or give remedial training to the retarded person. Again Harsanyi argues, that the maximin supporter would spend the money on the retarded person, while the Bayesian utilitarian would spend it on the mathematician.[46] Harsanyi believes that these two examples show that the Bayesian utilitarian would make the most reasonable decision. The use of the maximin principle, would lead, Harsanyi says, to unacceptable moral consequences.

The third argument that Harsanyi uses against the maximin principle in situations of uncertainty and in favor of the expected-utility maximization is not Bayesian. The argument focuses on what he calls the "equiprobability assumption."[47] Harsanyi maintains that decisionmakers ought to use this presupposition because it allows to treat all individuals' a priori interests as equal. According to Harsanyi, if there is no evidence showing that one event from a complete set of mutually exclusive events is more likely to occur than another, then individuals should judge the events equally probable.[48] For example, suppose that assessors are evaluating the probability that buried toxic chemical wastes will/will not migrate into a near aquifer within 10 years. If evaluators do not have evidence showing that one event (toxic wastes migrating into the aquifer) is more likely to occur than the other (toxic wastes will not migrate into the aquifer), then they should judge both events equally probable. Using the equiprobability assumption, together with expected-utility maximization as a decision rule under uncertainty, Harsanyi says, an individual always would choose the option that, in the person's opinion, would yield the higher average utility level to the members of the society.

---

45. See Harsanyi, Maximin, pp.40-43.

46. See Harsanyi, Maximin, pp.40-43.

47. See Harsanyi, Maximin, p. 45-46. See also Shrader-Frechette, *Risk*, pp. 112-16.

48. See Harsanyi, Maximin; and Shrader-Frechette, *Risk*, pp.104-16.

1.5.2. RAWLS AND THE CONTRACTARIAN STRATEGY

Proponents of the maximin principle, such as Rawls, maintain that in situations of uncertainty one ought to maximize the minimum; namely, one ought to avoid the decision having the worst possible consequences (given the possibility of ending in the worst-off position).[49] Rawls offers five main arguments to support the maximin strategy in situations of uncertainty. First, he argues that the maximin principle would lead to giving the interests of the worst-off the highest advantage. Second, Rawls maintains that the use of the maximin strategy would avoid employing a utility function, designed for risk taking, in the areas of morals, where it does not belong. Third, Rawls claims that the maximin principle would avoid the use of interpersonal comparisons of utility when defining justice. Fourth, he says it would avoid making supererogatory actions a matter of duty, as utilitarians theories do. Finally, Rawls argues that the maximin strategy would avoid the Bayesian/utilitarian dependence on uncertain predictions about impacts of alternative policies.[50]

Using his "original position," Rawls argues in favor of maximin and against expected-utility maximization. According to Rawls, we could arrive at just or fair social institutions if we were all rational individuals caring for our own interests, and if we bargained with each other about the nature of these institutions behind a "veil of ignorance." The veil of ignorance would prevent us from knowing our social positions, interests, abilities, or talents. Not knowing our own situation, Rawls says, we would arrange society so that even the worst-off individuals would not be seriously disadvantaged.[51] Thus, Rawls argues for the "difference principle," which is equivalent to maximin, because the principle proposes that one society is better than another if the worst-off members of the former do better that the worst-off members in the latter.[52] According to Rawls, the use of the maximin strategy under uncertainty

---

49. See Rawls, *Justice*; Rawls, Maximin; and Shrader-Frechette, *Risk*, pp. 116-26.

50. See *Shrader*-Frechette, *Risk*, pp. 116-26.

51. See Rawls, *Justice*; and Shrader-Frechette, *Risk*, pp. 116-23.

52. See, Rawls, *Justice*, pp. 75-83.

would prevent (where the expected-utility maximization principle might not) harming members of minority groups or disadvantaged people in order to increase the average utility of society.

A second argument in favor of maximin is that this principle would avoid using utility functions, designed for risk taking, in the area of morals where they do not belong. According to maximin supporters, utility judgments do not belong in the area of morals because they are based on preferences, while moral judgments are based on principles. Principles protect everyone, not only those who have free, well-informed preferences.[53] Maximin advocates, such as Rawls, maintain that utility functions express the subjective importance that we attribute to our own interests and needs, not the importance that we ought to attribute. For maximin supporters, then, equating preferences with "oughts" is questionable because we often prefer things that do not actually increase our welfare, such as smoking or drinking heavily. More generally, proponents of the maximin strategy maintain that, if our own welfare is identified in terms or our own preferences, then several undesirable consequences follow. First, we would ignore the quality of the choices and fall into moral relativism because all preferences would be assumed to be equally good. Therefore smoking and non smoking would appear as equivalent preferences. Second, our preferences would become inconsistent with the Bayesian/utilitarian assumption that tastes are stable, precise, and relevant to outcomes. Third, there would be no distinction between needs and wants, and between personal welfare and morality, because whatever we want would be defined as good. Fourth, we would assume that group welfare is merely the aggregate of individual preferences. Yet, proponents of maximin maintain that public well-being is not simply the aggregate of individual desires, because egoistic attitudes might serve each person's welfare but might destroy the common good.[54] Therefore, maximin advocates reject the use of utility functions in the area of morals.

A third argument in favor of maximin is that individuals would not have to make interpersonal comparisons of utility and therefore would not have to estimate their utility levels by placing themselves in the

---

53. See Shrader-Frechette, *Risk*, pp. 122-23.

54. See Rawls, *Justice;* and Shrader-Frechette, *Risk*, pp. 121-23.

specific situations of other individuals. Interpersonal comparisons of utility would then be avoided, maximin proponents say, because advocates of this decision rule can use merely an ordinal scale (that places far fewer burdens on the individuals attempting to evaluate consequences) rather that a Bayesian cardinal scale. Maximin supporters maintain that interpersonal comparisons of utility are difficult because different people's preferences do not have the same intensity, because stronger preferences are not always better, and because feelings of different individuals may not combine linearly.[55] Using maximin in situations of uncertainty, say its proponents, would avoid interpersonal comparisons of utility.

Maximin supporters also maintain that the use of the principle would avoid making supererogatory actions a matter of duty. If Bayesians and utilitarians were right, however, then one would always be required to perform both normal duties and heroic acts, because the Bayesian/utilitarian criterion for any action is whether it maximizes average utility. For example, people might be demanded under Bayesian and utilitarians rules to give up their own plans and instead to dedicate themselves to help all people dying of hunger. Such an obligation would present a problem in the light of our usual understanding of fairness and what is right and wrong.[56] Hence, maximin proponents claim that, in situations of uncertainty, one should not choose a Bayesian/utilitarian strategy.

A final argument in favor of the maximin principle is that it would avoid the Bayesian/utilitarian dependence on uncertain predictions about the effects of different policies. A consequence of this dependence is that two well-intentioned and well-informed Bayesians could each come to alternative decisions about what is right or wrong. For example, two Bayesian assessors evaluating commercial nuclear generation of electricity could arrive to different decisions. Thus, using the same evidence, one of the analysts could conclude that choosing commercial nuclear generation of electricity might cause catastrophic accidents in the future, while the other assessors could conclude that choosing commercial nuclear generation of electricity might cause no catastrophic

---

55. See Rawls, *Justice;* and Shrader-Frechette, *Risk*, pp. 123-24.

56. See Shrader-Frechette, *Risk*, pp. 124-25

accidents in the future. This difficulty is in part a result of the fact that many variables affect outcomes, and these variables are uncertain. Thus a Bayesian/utilitarian strategy is questionable because it relies on individuals' particular abilities to predict consequences.[57]

The next section argues that IVF assessors have chosen criteria for decisions under uncertainty consistent with expected-utility maximization rather than with the maximin rule. Similarly, they have preferred to minimize false positives over false negatives. However, because they are faced with a situation of uncertainty (about IVF) with potentially dangerous health and safety consequences, their decisions may foster public policies that underestimate the possibility of jeopardizing women's well being.

1.6. Making Ethical Decisions under Uncertainty

Science is frequently uncertain. For example, in quantum mechanics, physicists cannot always simultaneously determine the position and momentum of a subatomic particle. If science is sometimes uncertain, certainty obviously is not always required before people act. Thus, reasonable assurance about information and models, and nor certainty, is a prerequisite for ethically defensible behavior. What behavior is ethically justifiable under conditions of uncertainty (for example, about risks of particular pesticides) with potentially dangerous consequences? Should assessors be cautious and follow a maximin strategy that takes into account worst-case situations, or should they ignore such cases as highly unlikely? In this section I argue that IVF assessors should have used maximin rather than expected utility to evaluate IVF hazards, because a maximin approach would more likely protect women undergoing IVF against worst case situations such as cancer induction.

1.6.1. EXPECTED-UTILITY MAXIMIZATION VERSUS MAXIMIN

The case of IVF is one of uncertainty because of our total or partial ignorance about whether using this technique will result in a given outcome, such as having cancer, with a specific probability. Thus,

---

57. J. Rawls, Justice; and Shrader-Frechette, Risk, pp. 125-26.

assessors of IVF must decide whether it is better to recommend public policies consistent with criteria that maximize expected utility or with criteria, such as maximin, that avoid the worst possible consequence. In the case of IVF the worst possible consequence is cancer for the women that undergo the procedure and for their children.

As section 3.1 noted, there is abundant evidence that associates the drugs used for IVF with reproductive cancers. There is also a significant lack of knowledge about other possible IVF hazards (i.e., genetic defects, drugs health effects) for women and their children. However, assessors have sanctioned the use of IVF. Evaluators do not seem to have considered a worst case such as cancer. They seem to have used expected-utility maximization criteria rather than maximin.

Obviously, if IVF risks are uncertain, assessors cannot know the specific probabilities assigned to an outcome such as having reproductive cancer. Nevertheless, evaluators seem to have assumed that the possibility of cancer is low. As indicated in section 3.2, none of the four reports mentions cancer as one of the possible health hazards caused by IVF. Thus, assessors may have underestimated the probability of such outcomes.

Moreover, evaluators of this technique emphasize the benefits associated with IVF, that is, they stress the importance of having children. The Victorian report, for example, maintains that infertility is for many people "a serious, even tragic, deprivation."[58] The British report asserts that "childlessness can be a source of stress even to those who have deliberately chosen it."[59] According to this assessment, "for those who long for children, the realization that they are unable to found a family can be shattering. It can disrupt their picture of the whole of their future lives."[60] Likewise, the Spanish report affirms that "children are always a human hope, the fulfillment of a vital project, the perpetuation of oneself. Not having them truncates a fundamental joy."[61] As for the United States assessment, it states that for some people "having children is an important feature of their life plans because of their

58. Victorian Report, p. 4.
59. Warnock Report, p. 8.
60. Warnock Report, p. 8.
61. Spanish Commission, p. 53.

own experiences as children; their desire to have a link with the future or for emotional or genetic longevity."[62]

Because assessors appear to have underestimated IVF risks, yet emphasized IVF benefits, they seem to have sanctioned this technique because of a conclusion based on maximizing expected-utility. Although no IVF report explicitly mentions a decision criterion, assessment emphasis on benefits suggests expected-utility has driven the conclusions. In apparently using expected-utility maximization as a decision rule for IVF evaluations in a situation of uncertainty, analysts have erred because they may promote public policies that jeopardize women's health and well being. In neglecting worst-case situations, such as cancer for women undergoing IVF and for their children, evaluators may have discouraged governments from funding more research on adverse health effects of IVF. Assessors could also have deterred policy makers from promoting stricter controls in the use of this technology. For example, because doctors using IVF do not see the technique as dangerous, IVF assessments have not encouraged people a longer period of time before trying IVF. Thus, women's health and that of their children's may be put in danger.

Had IVF assessors used a maximin strategy (taking into account worst cases such as cancer) they might have encouraged policies that protect women's and children's well-being. Governments could, for example, have required more investigations on IVF hazards before widespread implementation of this procedure. Such a requirement might prevent the repetition of catastrophic cases such as that of DES.

Using expected-utility maximization presupposes that experts are right when they estimate dangers in a situation of probabilistic uncertainty. Preferring to minimize false positives over false negatives makes the same presupposition. The next section argues that, because IVF has potentially dangerous health impacts, the decision to minimize false positives may foster public policies that underestimate the possibility of jeopardizing women's well being.

---

62. OTA, Infertility, p. 37-8.

1.6.2. FALSE POSITIVES AND FALSE NEGATIVES

Consider the case of evaluators who must assess the uncertain consequences of implementing a new technology, such as *in vitro* fertilization, to treat infertility. Because the decision may have serious social, economic, and political consequences, analysts must determine what particular criteria to use in order to know how to interpret the uncertain results. For instance, if they overemphasize the risks that this new technique may pose to women, the community (e.g., industry and those who are infertile) may bear serious losses because of governments' restricting use and expansion of IVF. If, on the other hand, assessors underemphasize the dangers, many women (those who use IVF) could suffer significant health problems or even death. Thus because of the uncertainties in scientific information, when analysts reach their conclusions about implementation or regulation of a new technology, they must decide whether minimizing false positives is better or worse than minimizing false negatives, when both are not possible.

In a situation of uncertainty, false positives (type-I errors) occur when one rejects a null hypothesis that is true; false negatives (type-II errors) occur when one fails to reject a null hypothesis that is false. (An example of a null hypothesis might be, "IVF does not pose a statistically-significant increased risk of death to women over a ten-year period after the use of the technique.")[63] In assessing technological impacts under conditions of uncertainty, assessors must then decide whether to run the risks of rejecting a true null hypothesis (not using or expanding IVF when it is really safe and effective), or run the risks of not rejecting a false null hypothesis (using or expanding IVF when it is really unsafe and inefficient). Thus in situations of uncertainty, if evaluators minimize false positives when analyzing IVF, they minimize the possibility of restricting a harmless technology. On the other hand, if they decide to minimize false negatives, they minimize the error of accepting a harmful procedure.

---

63. See, for example, Shrader-Frechette, *Risk*, ch. 9; C. F. Cranor, *Regulating Toxic Substances* (New York: Oxford University Press, 1993), ch. 1 (hereafter cited as Cranor, *Regulating*); and K. Shrader-Frechette, *Ethics of Scientific Research* (Maryland: Rowman and Littlefield, 1994), ch. 6 (hereafter cited as Shrader-Frechette, *Research*).

Under conditions of uncertainty, decreasing the risks of false positives might result in underregulation of IVF and related techniques and therefore may hurt women's health. To minimize risks of false negatives might, however, produce overregulation. This strategy may impose excessive costs on manufacturers of IVF, and may hurt infertile people who want to use the technology.[64]

Preferences for minimizing false positives in situations of uncertainty may arise for several reasons.[65] Minimizing false positives, for example, appears more consistent with scientific practice. Hypothesis-testing in science functions on the basis of limiting false positives or limiting rejections of a true null hypothesis. The justification for this strategy seems to be that keeping the chances of rejecting a true null hypothesis low, researchers try to ensure an increase in scientific knowledge. Attempting to minimize false positives under conditions of uncertainty is also a common practice in criminal cases. In the criminal law the jury must be sure beyond a reasonable doubt that a defendant is guilty before convicting him. This is the basis for the idea that it is better for ten guilty people to go free rather than for one innocent person to be wrongly convicted.

Similarly, preferences for minimizing false positives in situations of uncertainty may arise because technical experts almost always use widely accepted Bayesian decision rules based on expected utility and subjective probabilities rather than the maximin principle.[66] As a result, even if everybody agrees that the probability of a high-consequence effect is uncertain but low, utilizing a Bayesian decision rule usually produces a choice in favor of the low-probability, but potentially-catastrophic, technological impact. In the same case, however, employing a maximin decision rule typically produces a conclusion against an uncertain but

---

64. Shrader-Frechette, *Risk*, ch. 9; Cranor, *Regulating*; and Shrader-Frechette, *Research*, ch. 6. See also, C. W. Churchman, *Theory of Experimental Inference* (New York: Macmillan, 1947).

65. See, for example, Shrader-Frechette, *Risk*, ch. 9; and Cranor, *Regulating*, chs. 2, 4.

66. See, for example, Shrader-Frechette, *Risk*, ch. 8; and Shrader-Frechette, *Research*, ch. 6. See, also, Rawls, *Justice*; Harsanyi, Maximin; Harsanyi, Rational; Harsanyi, Advances. See, also, M. Resnick, *Choices* (Minneapolis: University of Minnesota Press, 1986), pp. 26-37.

potentially dangerous consequence. Assessors' sanction of IVF use seems to indicate a preference for maximizing expected-utility rather than for using maximin. Thus, although the case of IVF is one of uncertainty with potentially dangerous consequences, evaluators have condoned the extensive use and expansion of the technique.

In the next section, I argue that in a case of uncertainty, such as that of regarding consequences of employing IVF, there are *prima facie* grounds for minimizing false negatives because underestimating the risks of this procedure may have disastrous consequences for women's health and well being.

1.6.3. ARGUMENTS FOR MINIMIZING FALSE NEGATIVES

There are several reasons for holding that, in situations of scientific uncertainty such as that of implementing IVF, assessors have a *prima facie* duty to minimize false negatives.[67] First, protecting the public from serious harm usually takes precedence over enhancing its welfare. Second, groups that are especially vulnerable need special protection. I treat these reasons in order.

First, in general, protecting people justifies minimizing false judgments that a potentially damaging technology such as IVF is harmless. Most political theorists would agree that protecting the public from serious harm (e.g., cancer and death) takes precedence over enhancing its welfare (e.g., by permitting wide IVF development).[68] The justification (for preferring to prevent harm rather than to enhance welfare under conditions of uncertainty when both are not possible) is

---

67. See Shrader-Frechette, *Risk*, ch. 9; Cranor, *Regulating*; and Shrader-Frechette, *Research*, ch. 6.

68. See, J. Bentham, "Principles of Civil Code," in *The Works of Jeremy Bentham*, 1., ed. J. Bowring (New York: Russell and Russell, 1962) (hereafter cited as Bentham, Principles); W. Frankena, *Ethics*, (Englewood Cliffs, NJ: Prentice-Hall, 1973) (hereafter cited as Frankena, *Ethics*); M. A. Slote, "The Morality of Wealth," in *World Hunger and Moral Obligation*, eds., W. Aiken and H. LaFollette (Englewood Cliffs, NJ: Prentice-Hall, 1977); RCNRT, *Care*, ch. 15; P. Singer, *Practical Ethics* (Cambridge: Cambridge University Press, 1993); and T. L. Beauchamp and J. F. Childress, *Principles of Biomedical Ethics*, 4th ed. (New York: Oxford University Press, 1994) , chs. 4-5 (hereafter cited as Beauchamp and Childress, *Principles*).

that protecting from harm appears to be a necessary condition for enjoying other freedoms. Thus, most liberal theorists would agree that individuals are responsible for obtaining their particular enjoyments, whereas the main responsibility of governments should be protecting from harm.[69] Also, health-care professionals often invoke the Hippocratic maxim: "Above all do no harm," to exemplify the need to protect patients against harmful actions.

Protecting women against risks to their health posed by IVF is a case of protection from harm rather than an enhancement of welfare, because risks to health are harms. As we have seen, women who undergo the IVF procedure may be exposed to serious hazards such as cancer and death. Obviously, women have rights to bodily security. But assessors of IVF have underestimated the existing scientific evidence and the insufficiency of data on IVF risks. Studies also show that it is likely that infertility doctors underemphasize uncertainty regarding this procedure because, in general, physicians tend to disregard uncertainty in their practices.[70] Thus because women's lack of information may hinder their abilities to give free informed consent, using techniques that might cause cancer and death could jeopardize women's rights to bodily security.

Similarly, disease and disability, induced by IVF, may restrict the range of opportunities individuals have opened to them.[71] Illness also may limit appreciation of other enjoyments. Employing drugs and procedures that may imperil women's health may then disrupt their general well-being. Therefore, in underestimating the risks of a technology that may produce death and illness, assessors have failed to protect against loss of a good, namely women's health and well being, and thus they have failed to prevent harm. Someone might argue that restricting IVF is not a case of preventing harm because women want it, and choosers do not select to harm themselves. However, as I argue in

---

69. See, for example, Bentham, Principles; and Frankena, *Ethics*,

70. See, for example, J. Katz, "Why Doctors Don't Disclose Uncertainty," *The Hastings Center Report*, 14: 1 (1984): 35-44; RCNRT, *Care*, p. 545-48, and R. L. Logan, "Uncertainty in Clinical Practice: Implications for Quality and Costs of Health Care," *Lancet*, 347 (1996): 595-98.

71. See, for example, N. Daniels, *Just Health Care* (New York: Cambridge University Press, 1098).

chapter seven, because women lack information on IVF risks and benefits, we cannot say that they really choose to undergo IVF treatment. Therefore, restricting use of IVF is a case of protecting from harm, because women are not adequately informed.

Granted, permitting the implementation of IVF may constitute a case of enhancing welfare for some people (those who might be able to have children through the procedure.) However, because the success rate of IVF is quite low (around 10%), because cancer and death may be some of the outcomes, because the uncertainties about possible risks are great, and because there are other medical and preventive alternatives to IVF, then protecting women against risks to their health posed by IVF is more a case of protecting from harm than of enhancing welfare. Thus, if protecting from harm (such as illness and death) is more important than providing some good (such as having babies) then, in situations of uncertainty with potentially dangerous consequences, IVF assessors ought to minimize false negatives, rather than false positives, to the degree that minimizing false negatives prohibits positive harm, that is, protects from serious threats rather than enhances welfare.

A second reason for limiting false negatives in situations of uncertainty, when evaluating IVF, is that because women historically have been a vulnerable group, they typically need protection. Even when women decide to undergo IVF treatment, they may be helpless because (as I argue in chapter seven) they lack adequate information on IVF risks and benefits.

Women are vulnerable when facing medical problems and need more protection because usually they lack financial resources (they have less access to the job market than men and are often more financially dependent than men). They also are vulnerable because frequently they lack information to deal with the dangers that may affect them, as cases such as the implementation of DES, the Dalkon Shield, and Thalidomide show.[72] Likewise, women are especially vulnerable because of the pressure that society imposes on them to become mothers.[73] Because of

---

72. See Shrader-Frechette, *Risk*, ch. 9;Beauchamp and Childress, *Principles*, chs. 4-5; and Shrader-Frechette, *Research*, ch. 6.

73. See, for example, R. Arditti, R. D. Klein, and S. Minden, *Test-Tube Women. What Future for Motherhood* (London: Pandora Press, 1984); B. K. Rothman,

this expectation, infertility generates a grave problem for women because they often accept the belief that childbearing is necessary for a satisfactory life. As a consequence, they are particularly vulnerable, for they are less likely to be concerned about risks to their lives if they think they will have a chance of becoming mothers.

Women's vulnerability and their need for protection is also apparent when observing unjust discrimination against women in the provision of health care and in health research. For example, recent studies indicate that when women experience renal failure, they receive fewer kidney transplants than men. Females between the ages of forty-six and sixty are only half as likely to receive a transplant as males of the same age.[74] Similarly, a study completed in 1987 showed that, all things being equal, men were 6.5 times as likely to be referred for cardiac catheterization (a prerequisite for coronary bypass surgery) than women, although men have only three times the likelihood of having coronary heart disease.[75] Medical research practices such as involuntary sterilization, hysterectomy, mastectomy, and high rates for cesarean births also show the necessity to protect women.[76] Women are also particularly vulnerable because of the predominance of males in the biomedical sciences. Medicine has historically excluded women and, still today, women constitute a minority in the hierarchy of this profession. Such exclusion may increase the chances of patronizing, unperceptive, and harmful attitudes toward female patients, including those using IVF.[77]

Recreating Motherhood. Ideology and Technology in a Patriarchal Society (New York: W.W. Norton & Company, 1990); A. Phoenix, A. Woollett, and E. Lloyd, eds., *Motherhood. Meanings, Practices, and Ideologies* (London: Sage, 1991); M. S. Ireland, *Reconceiving Women* (New York: The Guilford Press, 1993); and R. Jackson, *Mothers Who Leave* (London: Pandora, 1994).

74. See H. and J. Lindemann Nelson, Justice, p. 359.

75. See, J. N. Tobin *et al.*, "Sex Bias in Considering Coronary Bypass Surgery," *Annals of International Medicine* 107 (1987): 19-25; R. M. Steingart *et al.*, "Sex Differences in the Management of Coronary Artery Disease," *New England Journal of Medicine* 325 (1991): 226-30; and H. and J. Lindemann Nelson, Justice, p.359.

76. See, for example, Committee on Labor and Human Resources, *Women's Health: Ensuring Quality and Equity in Biomedical Research* (Washington, DC.: U.S. Government Printing Office, 1992); and Mastronianni *et al.*, eds., *Women and Health Research.*

77. See, for example, M. A. Warren, "Is IVF Research a Threat to Women's

Because of the vulnerability of women in relation to medical issues in general, and reproductive problems in particular, IVF assessors have a *prima facie* duty to try to protect them from harms associated with this procedure. Therefore, evaluators have a *prima facie* duty to limit false negatives, given that doing so would likely protect women.

If assessors have a *prima facie* duty to minimize false negatives under conditions of uncertainty, then evaluators' decision to minimize false positives (in a situation of uncertainty having potentially dangerous consequences) is questionable. This preference is problematic because evaluators may have encouraged public policies that underestimate the possibilities of jeopardizing women's heath. For example, because decisionmakers might think that IVF does not pose serious risks for women, they might refrain from implementing stricter regulations in the application of IVF. Similarly, because they might assume that IVF is a safe treatment, policymakers might avoid or restrict funding for research about IVF as an infertility therapy.

1.7. Sanctioning the Expansion of In Vitro Fertilization

IVF assessors also have implicitly sanctioned questionable criteria for making decisions under uncertainty when condoning the expansion of IVF to an increasing number of reproductive conditions. Initially, doctors intended IVF only for women with reproductive problems caused by blockage of the fallopian tubes. Now professionals use IVF in cases of endometriosis, unexplained infertility, sperm antibodies, cervical factors, and male-related conditions, such as low sperm count. Also, clinics employ IVF for postmenopausal women, for couples who need donor gametes, for the formation of embryos to donate to another couple, or to transfer to a surrogate for gestation when a woman is unable to continue a pregnancy.[78]

Autonomy?" in Embryo *Experimentation*, eds., P. Singer *et al.*, (Cambridge: Cambridge University Press, 1990), pp. 125-40.

78. See, for example, P. M. Lewis, "Patient Selection and Management," in *A Textbook of in Vitro Fertilization and Assisted Reproductive Technology*, eds., Brinsden and Rainsbury, pp. 27-37; K. Dawson, "Ethical Aspects of IVF and Human

Assessors have argued that, given the importance that people place on having children, expanding use of IVF to conditions other than tubal occlusion is acceptable because doctors could help more people. The Victorian committee, for example, claims that although patients with tubal disease are the most common group to whom doctors have applied IVF, couples with unexplained infertility of prolonged duration and those with reproductive problems due to a disorder of sperm production also could profit from it.[79] The British Commission also refers to the use of IVF as a treatment for oligospermia (scarcity of sperm in the semen) and unexplained infertility.[80] The Spanish committee mentions tubal disease, endometriosis, ovarian problems, male factors (i.e., low-sperm count), unexplained infertility, and the prevention of genetic diseases as indications that could benefit from IVF.[81] The United States report says that practitioners utilize IVF in a number of disorders that include tubal disease, endometriosis, idiopathic infertility, cervical mucus abnormalities, oligospermia, and any combination of these disorders.[82]

As we have seen in sections two and three of this chapter, because assessors have underestimated both the existing scientific evidence and the insufficiency of data on IVF hazards, sanctioning the application of IVF to an increasing number of reproductive problems is questionable. More women may undergo risky and unnecessary treatment when they are ignorant of the known hazards and they are not fully aware that the procedure may have unidentified dangers.

Condoning doctors' recommendations to expand the use of IVF also is questionable in light of the insufficiency of data on the success of the treatment for new conditions such as unexplained or male-related infertility. Because IVF may be effective for certain indications but not

---

Embryo Research," in *Handbook of in Vitro Fertilization*, eds., A. Trounson and D. K. Gardner (Boca Raton: CRC Press, 1993), pp. 287-302; M. D. Damewood, "In Vitro Fertilization and Assisted Reproductive Technologies," in *Reproductive Medicine and Surgery*, eds., Wallach and Zacur, pp. 845-859; S. H. Chen and E. E. Wallach, "Five Decades of Progress in Management of the Infertile Couple," *Fertility and Sterility*, 62:4 (1994): 665-685;

79. Victorian Report, p.11.

80. Warnock Report, p.29.

81. Spanish Commission, p. 106-107.

82. OTA, *Infertility*, p.123.

for others, it is essential to determine the effectiveness of IVF for each condition. To determine efficacy, investigators might use two different methods: animal research and human randomized clinical trials (where doctors allocate patients to treatment with IVF, to other treatments for infertility, and to no treatment at all). As for animal research, there is little available data on IVF efficiency. Similarly, the existent randomized clinical trials on IVF are inadequate because of methodological weaknesses. For example, small sample sizes, even when researchers aggregate study results to do meta-analysis, limit the conclusions that they can draw about IVF effectiveness. According to some investigations, few randomized clinical trials are of sufficient quality even to allow meta-analysis (that is, pooling the results from studies with similar methodologies so as to consider larger sample sizes to provide reliable results). For example, some of these studies do not state the method of randomization, or they do not have any control ("no treatment") groups.[83] The data that would allow researchers to examine fully this technique for effectiveness is, therefore, incomplete and controversial. Assessors of IVF, nevertheless, seem to have underestimated the limited evidence.

In condoning the expansion of the use of IVF to conditions other than tubal occlusion, and lacking adequate evidence on risks and effectiveness, assessors again have preferred to minimize false positives over false negatives under conditions of uncertainty. In doing so, they have failed to prevent harm. However, if my previous arguments for the *prima facie* duty of IVF evaluators to prevent harm are correct, then their decision is questionable. In what follows I show that protecting women against expansion of IVF to conditions others than tubal blockage also is a case of protecting from harm.

Protecting women against expansion of IVF is a case of protecting from harm rather than enhancing welfare because more women would be exposed to risks of cancer and death if they used IVF. As I have argued in the previous section of this chapter, protecting women against risks to their health posed by IVF is a case of protecting from harm because, even though women may choose to undergo IVF treatment (and they would not choose harm) they do not have adequate information on IVF hazards

83. See, RCNRT, *Care*.

and benefits. Thus, limiting IVF is a case of protecting from harm because those who undergo this procedure may be subject to serious dangers such as cancer and death that may be unknown to them. Condoning the expansion of IVF would increase the number of women who could be exposed to these hazards. Exposing more women to the risks associated with hormones and invasive treatments (without their consent) might, as I have argued in section six, undermine women's rights to bodily security. It also might restrict the range of opportunities (because of illness or disability) individuals have open to them. Therefore, protecting women against such an expansion would be a case of protection from harm.

Again, warranting the expansion of IVF may enhance the welfare of some people who might be able to conceive a child. Nevertheless, because IVF has not been proved effective for conditions other than tubal occlusion, because its success rate is low, because of the seriousness of some IVF outcomes (cancer), and because there are other options for childless people, protecting women against expansion of IVF is more a case of protecting from harm than one of enhancing welfare.

If protecting women against expansion of IVF is a case of protecting from harm, and if protecting from harm usually takes precedent over enhancing welfare, then assessors have erred. They have erred because, faced with a situation of uncertainty with potentially dangerous consequences, they have preferred to minimize false positives. In doing so, assessors might have encouraged public policies that underestimate the possibility of endangering women's health. Evaluators' preferences for false negatives might foster policy decisions that put women at risk in a variety of ways. First, and most obviously, following IVF assessments, decisionmakers might condone the use of IVF for conditions other than tubal occlusion without their having adequate evidence concerning safety and efficiency. Hence, many women may undergo IVF treatment and be exposed to serious health risks when there is no evidence that it would help them to conceive a child. They may, therefore, be exposed to the hazards of this technology without knowing whether they are more likely to have a child than if they received no treatment or an alternative one.[84] And second, decisionmakers might select public policies that put women

84. See RCNRT, *Care*

at risk because, in presupposing with the assessors that the evidence on IVF effectiveness (for conditions others than tubal occlusion) is adequate, policymakers might avoid or limit funding for research on IVF as an infertility therapy.

1.8. Some Objections and Responses

There are several objections that critics may pose to my analysis of IVF assessments. First, they may argue that evaluators have recommended obtaining written informed consent from women as a way to overcome problems with risks. Second, critics may claim that assessors' sanction of the use and expansion of IVF under conditions of uncertainty is adequate because restricting the use of this procedure would interfere with women's abilities to choose a risky technology, and such interference would be paternalistic. I treat these two objections in order.

1.8.1. FREE INFORMED CONSENT AS A WAY TO OVERCOME PROBLEMS WITH RISKS

The first criticism of my analysis of IVF assessments (that evaluators have recommended obtaining written informed consent from women as a way to overcome problems with risks) seems particularly compelling because it emphasizes the requirement of informed consent. I argue, however, that this objection is questionable because it ignores the fact that if women lack adequate information about IVF risks and benefits, then they cannot give genuinely informed consent. In chapter seven, I offer a more complete account of problems with IVF evaluations that might hinder women's opportunities to give free informed consent.

As I discuss in chapter seven, legal, regulatory, medical, psychological, and philosophical literature tend to evaluate informed consent in terms of the following analytical components: (1) disclosure; (2) understanding; (3) voluntariness; and (4) competence.[85] Scholars

---

85. See, for example, A., M., and L. H. Roth, "What We Do and Do Not Know About Informed Consent," *Journal of the American Medical Association*, 246:21 (1981): 2473-2477; U.S. President's Commission for the Study of the Ethical

argue that one gives informed consent to an intervention if and only if one is competent to act, receives a thorough disclosure about the procedure, understands the disclosure, acts voluntarily, and consents to the intervention. Disclosure, the main component of consent (from an institutional point of view), refers to the necessity of professionals' passing on information to decisionmakers and possible risk victims. Understanding, the second element in the process of obtaining free informed consent, requires professionals to help potential risk victims overcome illness, irrationality, immaturity, distorted information, or other factors that can limit their grasp of the situation to which they have the right to give or withhold consent. Voluntariness, or being free to act in giving consent, requires that the subjects act in a way that is free from manipulation and coercion by other persons. Finally, the criterion of competence demands that potential risk victims give autonomous authorization to some act, such as approving certain fertility treatment.

Given these elements that scholars recognize as necessary for informed consent, to claim that women can give free informed consent to IVF treatment is questionable. If women are ignorant of the fact that IVF is of unproved benefit for many infertility conditions, and if they are not aware that the treatment may have unknown risks, then they cannot give genuinely informed consent. Lack of information may seriously hinder people's abilities to make informed choices. Some of the relevant data necessary for insuring free informed consent of women undergoing IVF include (1) personal chances of having a child as a result of the treatment; (2) alternatives to the procedure; (3) long-term effects; (4) emotional demands that the treatment imposes (timing of sexual relations, frequent visits to the clinic, failing to achieve a pregnancy); and (5) short-term consequences of the treatment. According to some recent studies, the

---

Problems in Medicine and Biomedical and Behavioral Research, *Making Health Care Decisions: A Report on the Ethical and Legal Implications of Informed Consent in the Patient-Practitioner Relationship* (Washington, D.C.: U.S. Government Printing Office, 1982); R. Faden and T. Beauchamp, *A History and Theory of Informed Consent* (New York: Oxford University Press, 1986); ch. 5; and Beauchamp and Childress, *Principles*, ch. 3.

majority of IVF patients surveyed were dissatisfied with the information received in these areas.[86]

Critics still may argue that it is the individual's responsibility, and not that of the assessors', to inform themselves about IVF risks and benefits as fully as possible. IVF evaluators' roles are not to educate infertile people on the hazards and effectiveness of this procedure but to analyze the data they consider relevant for policymaking. This objection, however, is questionable. Because government IVF commissions are fundamental steps in the adoption and implementation of particular policy decisions, they have an obligation to ensure that inappropriate or unethical use of IVF is prohibited and that the necessary mechanisms to protect informed decisionmaking are in place. Furthermore, given the infertile couples' strong desire to utilize all possible successful treatments, their ability to give genuinely informed consent could be compromised. Because of the committees' influence on the approval of certain public policies, they also have a responsibility to reduce, where possible, vulnerability in relation to infertility treatments and to ensure that those who are in positions of power and authority do not manipulate or control those who are vulnerable. For this reason, accurate information seems particularly important.

1.8.2. INTERFERING WITH WOMEM'S RIGHTS TO CHOOSE RISKY TECHNOLOGIES

A second objection to my analysis of IVF assessments is that evaluators' sanctioning the use and expansion of IVF under conditions of uncertainty is adequate because restricting the use of this procedure would interfere with women's abilities to choose a risky technology whose benefits they desire. According to those who support procreative rights, such interference would be paternalistic.[87] I argue that criticizing my analysis for encouraging interference with the implementation and use of IVF, on grounds of unjustified paternalism, fails. The objection errs because such interference may not be paternalistic at all or may be a case of justified

---

86. See RCNRT, *Care*, p. 547.

87. See, for example, J. Robertson, *Children of Choice* (Princeton: Princeton University Press, 1994) (hereafter cited as Robertson, *Choice*).

paternalism for several reasons. First, women's consent to IVF is not adequately informed. Second, extensive use of IVF may harm other members of the community such as children, the poor, and minority women. I treat these reasons in order.

Philosophers often define 'paternalism' as the intentional overriding of one person's preferences or actions by another individual or institution under the justification of benefiting or avoiding harm to the person whose will is overridden.[88] Most moral and political philosophers also agree that paternalistic interventions are sometimes justified.[89] Similarly, they usually distinguish between weak (soft) and strong (hard) paternalism.[90] In weak paternalism, an agent intervenes on grounds of beneficence or nonmaleficence only to protect people against their own substantially nonautonomous actions. These actions include cases of consent that are not adequately informed, severe depression that precludes rational deliberation, and critical addiction that precludes free choice and action. Strong paternalism, on the other hand, involves interventions intended to prevent harm to an individual despite the fact that the individual's risky actions are informed, voluntary, and autonomous.[91] The case of IVF constitutes, as we shall see, a case of weak paternalism.

If the brief discussion of informed consent in section 8.1 is correct, then we cannot say that women's consent to IVF is adequately informed, because many women appear to lack relevant information on IVF risks and benefits. (See chapter seven for a more complete account of problems that may obstruct women's abilities to give free informed

---

88. See, for example, J. Feinberg, "Legal Paternalism," *Canadian Journal of Philosophy* 1 (1971): 105-24 (hereafter cited as Feinberg, Paternalism); G. Dworkin, "Paternalism," *The Monist* 56 (1972): 64-84; I. Kant, *On the Old Saw: That May Be Right in Theory But It Won't Work in Practice*, trans. E.B. Ashton (Philadelphia: University of Pennsylvania Press, 1974); J. S. Mill, *On Liberty. Collected Works of John Stuart Mill*, vol. 18 (Toronto: University of Toronto Press, 1977); J. Kleinig, *Paternalism* (Totowa, NJ: Rowman and Allanheld, 1983); D. Van DeVeer, *Paternalistic Intervention* (Princeton, NJ: Princeton University Press, 1986); and Beauchamp and Childress, *Principles*, pp. 271-291.

89. See previous note for references.

90. See, for example, Feinberg, Paternalism, 105-124.

91. See Feinberg, Paternalism, pp. 113, 116; and Beauchamp and Childress, *Principles*, pp.277.

consent). Therefore, promoting restrictions on the use of IVF would be a case of weak paternalism because women cannot give free informed consent to the procedure. But, that persons deserve to be protected from harm caused to them by conditions beyond their control is hardly questionable.[92] If this is so, then evaluators' support for regulation or restriction of IVF does not constitute a case of strong or unjustified paternalism, because women are adequately informed neither about the possible hazards to their health nor about the likelihood of becoming pregnant through this procedure.

Second, encouraging interference with women's procreative liberty by restricting use of IVF may not be paternalistic in the strong sense because such use may harm others. Certainly, procreative liberty is extremely important because of the effects that the decision to have or to refrain from having children may present to our sense of dignity, identity, and meaning of life.[93] However, emphasizing the primacy of procreative liberty risks overlooking the fact that reproduction clearly involves the community.[94] Reproduction always occurs with a partner and normally requires the collaboration of doctors and nurses. Moreover, reproduction affects others by creating a new person. In centering the debate on the right to have children, critics may treat as irrelevant to moral analysis or public policy the effects of reproductive choices on women, on offspring, on family, and on society.

Furthermore, restricting IVF and related technologies may not be paternalistic in the strong sense because these procedures also can affect other members of the community and the community as a whole by changing profoundly held values. For example, using IVF techniques, a child may have up to three mothers: the woman who provides the egg, the woman who bears the child, and the one who rears her. The impacts of IVF on children's psychological well being and on society in general

---

92. See, J. Feinberg, *Harm to Self*, vol. III of *The Moral Limits of the Criminal Law* (New York: Oxford University Press, 1986), p. 14; and Beauchamp and Childress, *Principles*, pp.277.

93. See, for example, Robertson, *Choice*.

94. See, for example, M. A. Glendon, *Rights Talk: The Impoverishment of Political Discourse* (New York: Free Press, 1991); E. Kingdom, *What's Wrong with Rights: Problems for Feminists Politics of Law* (Edinburgh, U.K.: Edinburgh University Press, 1991).

are unforeseen and may be far-reaching. Also, the use of IVF may harm women as a social group. For example, these medical procedures affect women as a group by promoting commercialization of reproduction. Sperm, eggs, embryos, and babies now have become commodities. This is obvious when the donation of gametes or the use of the uterus (in so-called "surrogate motherhood") involves profits for the donor. Exploitation of low-income women would be a clear reality in a system where gametes, embryos or wombs can be bought and sold. Participants at the UNESCO 1985 International Symposium (on the Effects on Human Rights of Recent Advances in Science and Technology) pointed out the problem of social and racial discrimination in the case of reproductive medicine. The Conference recognized that those most at risk from the assisted-conception procedures are poor, migrant, refugee, or ethnic minority women. Such risk comes, for example, from the possibility of using them as "Guinea pigs" in research on these technologies, as producers of eggs, or as surrogate mothers.[95]

If use of IVF may substantially harm other members of the community, or women as a whole, then restrictions on this procedure may not be paternalistic at all or may be justified paternalistic actions. Certainly, the social consequences of the extensive utilization of IVF and related technologies are far from known. However, when assessing these procedures, evaluators also should analyze the ways in which reproductive decisions may alter and reflect all kinds of social forces. Total freedom to choose a technological option such as IVF may damage the achievements of liberty for many women, for people living in poverty, or for disabled people.

Furthermore, in countries with socialized health care such as Australia, Spain, and the United Kingdom, assessors charged with

---

95. See UNESCO, *International Symposium on the Effects on Human Rights of Recent Advances in Science and Technology* (Paris: Division of Human Rights, 1985). See, also, Rowland, *Laboratories,* 211-216; Koval, R., "The Commercialization of Reproductive Technology," in *Baby Machine. Reproductive technology and the Commercialization of Motherhood,* ed., J. A. Scutt (Melbourne: McCulloch Publishing, 1988), pp. 108-134; and Corea, G., "Women, Class, and Genetic Engineering. The Effect of New Reproductive technologies on all Women," in *Baby Machine. Reproductive technology and the Commercialization of Motherhood,* ed., Scutt, pp. 135-156.

evaluating medical technologies have a duty to ensure that their evaluations do not encourage inefficient use of scarce resources. Not to do so may harm the public. Because the state's money is always scarce, decisions about resource allocation may have important ethical, economic, and political consequences. Encouraging public policies that use taxpayers money on an inefficient and expensive technology such as IVF might prevent decisionmakers from spending those same resources on more beneficial measures.

1.9. Summary and Conclusion

Because the Victorian, the British, the Spanish, and the United States reports influence public policies in relation to infertility treatments, evaluating their deficiencies is extremely important. Without such assessments, governments may make decisions that are not in the public's best interests. This chapter has analyzed some of the inadequacies present in the reports mentioned. I have argued here that IVF assessors have underestimated the possibility of jeopardizing women's health because they have neglected epistemological and ethical problems such as choosing criteria for decisions under uncertainty. I have defended the thesis that evaluators have erred in their conclusions because, in a situation of uncertainty with potentially dangerous consequences, they have condoned the use of IVF. Thus, they implicitly have sanctioned questionable criteria, such as expected-utility maximization, for deciding in a situation of scientific uncertainty. They also have preferred to minimize false positives over false negatives. As a consequence, decisionmakers might promote policies that underestimate the possibility of endangering women's health. Likewise, I have claimed that, in condoning the expansion of IVF to an increasing number of reproductive problems, assessors also implicitly have endorsed questionable criteria for making decisions under scientific uncertainty. In doing so, they may be responsible for persuading many women to undergo useless treatments and be exposed to unnecessary dangers associated with risky drugs and invasive procedures.

CHAPTER 5

IN VITRO FERTILIZATION AND THE ETHICS OF NONCONSERVATIVE MEDICAL CARE

1.1. Introduction

In 1991, the United States spent 13.2% --$751.8 billion-- of its gross domestic product (GDP) on health care.[1] The expenditure curve is still rising, with $884.2 billion in 1993.[2] Despite the amount devoted to health expenditures, over 14% of people living in the United States have no health insurance, and an even greater percentage are seriously underinsured.[3] Although researchers disagree about how much technology has contributed to increasing health care costs, they generally agree that medical technology is one of the main reasons for this increment.[4] Around 20% of health costs are spent on procedures and services that are, in many cases, clearly unnecessary such as caesarian sections, hysterectomies, laminectomies (back surgery), or diagnoses with magnetic resonance imaging.[5]

Our society tends to believe that more technology is always better.[6] This heavy dependency on technology (especially nonconservative techniques), to

---

1. See, H. Lindemann Nelson and J. Lindemann Nelson, "Justice in the Allocation of Health Care Resources: A Feminist Account," in *Feminism and Bioethics*, ed., S.M. Wolf (New York: Oxford University Press, 1996), pp. 351 (hereafter cited as H. and L. Lindemann Nelson, Justice).

2. See, S. M. Ayres, *Health Care in the United States* (Chicago: American Library Association, 1996), p.2.

3. See H. and L. Lindemann Nelson, Justice, p.351.

4. See, for example, P. J. Neumann and M. C. Weinstein, "The Diffusion of New Technology: Costs and Benefits to Health Care," in *The Changing Economics of Medical Technology*, eds., A. C. Gelijns and E. A. Halm (Washington, D.C.: National Academy Press, 1991), pp. 21-34.

5. See "Wasted Health Care Dollars," *Consumer Reports*, 57: 7 (July 1992): 435-448 (hereafter cited as Consumer Reports).

6. See, for example, D. Mechanic, *From Advocacy to Allocation: The Evolving*

fix our health problems, at the exclusion of more conservative solutions, is also tied to a preference to look always for the easiest--even if not the wisest--answers to our difficulties. The quest for increasingly higher levels of techniques in order to avoid more problematic changes in lifestyle is clear even when it becomes evident that they cannot solve our problems.

Neonatal intensive care in the U.S. constitutes an example of this trend to develop new technologies rather than promote institutional changes that could save more children's lives. The annual financial costs of neonatal intensive care in the U.S. vary from $2.8 to $4 billion. If the incidence of low birthweight were reduced by 1% the U.S. could save over $400 millions a year in immediate neonatal intensive care charges.[7] We know that the adverse effects of low socioeconomic status are important determinants of child health.[8] Thus, improving the social and economic conditions of mothers and children in need and promoting comprehensive pre-conception, prenatal, and infant care could be a cost-effective solution toward reducing the low-birthweight, and infant mortality rates.[9]

1.1.1. OVERVIEW

The same apparent asymmetry (deciding to spend thousands of dollars on new technologies and ignoring social changes and preventive measures that could better solve the same problems) appears in the evaluation of *in vitro* fertilization (IVF) and related procedures. The goal of this chapter is to argue

---

American Health Care System (New York: Free Press, 1986); R.H. Blank and M.K. Mills, *Biomedical Technology and Public Policy* (New York: Greenwood Press, 1989).

7. See Robert M. Kliegman, "Neonatal Technology, perinatal Survival, Social Consequences, and the Perinatal Paradox," *American Journal of Public Health* 85: 7 (1995): 109-13 (hereafter cited as Kliegman, Neonatal). See, also, R. A. Rosenblatt, "The Perinatal Paradox: Doing More and Accomplishing Less, "*Health Affairs*, 8 (1989): 158-168.

8. See, for example, C. J. R. Hoghe and M. A. Hargraves, "Class, Race, and Infant Mortality in the United States," *American Journal of Public Health* 83 (1993): 9-12; and Kliegman, Neonatal

9. See, for example, A. Racine, T. Joyce, and M. Grossman, "Effectiveness of Health Care Services for Pregnant Women and Infants," *The Future of Children*, 2 (1992): 40-55; and Kliegman, Neonatal.

that IVF assessments are problematic because they might have encouraged public policies that pay too little attention to the common good. In section two, I offer a brief overview of theories that emphasize individual rights as opposed to those that stress the common good. I also present an account of how IVF evaluators have addressed individual rights and common good issues. In section three, I argue that evaluators of IVF have erred in their analyses because, in presenting the problem of infertility as primarily an individual one, they have underestimated the role of social, ethical, and political solutions, such as prevention, in solving reproductive problems. Undervaluing these kinds of solutions is, however, problematic because it may underemphasize social influences on, and the community's responsibility for, the well being of its members and may foster unfair discrimination against women. In section four, I argue that IVF assessors have failed in their analyses because, in assuming too broad a definition of 'infertility', they may encourage people to use IVF needlessly. Thus, assessors seem to have conceded more importance to individuals' desires to have genetically related children than to problems of the common good. However, unnecessary use of IVF may affect the common good because it might influence governments to increase needlessly the amount of resources dedicated to this technique. Promoting excessive funding of IVF is problematic because IVF and related technologies are expensive procedures. Given that health resources are scarce, allocating money for IVF would likely prevent the funding of other health-care measures that might be equally or more important.

I also try to answer some possible objections to my arguments. First, critics may emphasize the difficulty of showing that social, ethical, and political solutions are cost-effective. Second, they may stress the fact that these kinds of answers to reproductive problems may require unattainable institutional changes. Third, critics also may object that the definition of 'infertility' that assessors have used more likely would enhance the welfare of more individuals. Fourth, they may argue that IVF evaluators have used the best available information on the prevalence of infertility. Finally, critics may claim that IVF analysts have adopted the standard medical definition of infertility.

1.1.2. SOME CAVEATS

Some clarifications are in order before developing my arguments. First, I define 'conservative solutions' to the problem of infertility as those options, such as preventive measures, that often are less risky, have lower costs (at least in the long-term), and are more accessible to people. Second, I shall not debate here whether solving infertility problems is worthy of public support. I shall assume, as the reports in general do, that it is. Two of the assessments -- those of the British and the Spanish Commissions-- specifically recommended that IVF be funded, at least in part, by public money. My focus is the role that the reports have conceded to IVF in the alleviation of reproductive problems and the consequences of their analyses for our society. Third, I am not arguing that public and private funding of IVF therapy and research should stop. Mine is an argument about priorities. Discussions about IVF and related technologies often have neglected to consider the economic restrictions prevalent in most health-care systems today. No health system can afford to do everything. In order to maximize public health in the face of these constraints, countries have to allocate health-care resources in the most equitable way possible and with maximum benefit to the public's health.[10] I think that often, conservative medical care is a more appropriate choice to fulfill those goals. But I am not defending the thesis that nonconservative solutions, such as IVF and related technologies, should be excluded from public funding. Fourth, I am not defending the thesis that an individual-rights approach is inherently harmful for the community. Although rights-based theorists defend the thesis that individual rights (particularly negative rights) should rule unless there are compelling reasons to the contrary and that social claims are seldom sufficient to limit individual rights, rights have played a vital historical role in the evolution of

---

10. See, for example, N. Daniels, *Just Health Care* (Cambridge, England: Cambridge University Press, 1985); P.-G. Svensson and P. Stephenson, "Equity and Resource Distribution in Infertility Care," in *Tough Choices*, P. Stephenson and M. G. Wagner, eds., (Philadelphia: Temple University Press, 1993), pp. 161-166; T. L. Beauchamp and J. F. Childress, *Principles of Biomedical Ethics* (New York: Oxford University Press, 1994), ch.6; T. A. Mappes and D. DeGracia, *Biomedical Ethics* (New York: McGraw-Hill, 1996), ch. 10; and H. and L. Lindemann Nelson, Justice; E. H. Loewy, *Textbook of Healthcare Ethics* (New York: Plenum Press, 1996), chs. 2,11.

our society. Many of those rights, i.e., free speech, due process, make communal deliberation and democracy possible. My point, then, is not to defeat the importance of rights, but to emphasize that stressing the importance of individual rights may blind assessors to harmful consequences of IVF technology for the common good.

1.2. Individualism and the Common Good

The relationships among individuals and their communities have concerned ancient philosophers such as Socrates and Aristotle, medieval scholars such as Thomas Aquinas, modern theorists such as Hobbes and Locke and contemporary authors such as Rawls and MacIntyre. In this section I offer a brief overview of what has been a continuous debate in ethical and political philosophy, namely the discussion between individualists and supporters of the common good. Next, I argue that assessors' analyses of IVF and related procedures have presented the problem of infertility mainly as an individual-choice issue and have neglected the possible impacts of these technologies on the community.

1.2.1. INDIVIDUALS VS. COMMUNITIES: THE THEORY

The use of metaphors in science, medicine, and philosophy is frequent. Images of animals' bodies as machines, the earth as a lifeboat, the immune system as a battleground, or cells as factories for conversion of energy constitute some examples.[11] Some of these metaphors are permeated with social values and shape our expectations about scientific theories, political conceptions, and ethical views. For example, medical books describe female reproductive features as a machine whose main function is reproduction.[12] In this image, menstruation appears as failed production, and menopause as an

11. See, for example, R. J. Sternberg, *Metaphors of Mind: Conceptions of the Nature of Intelligence* (Cambridge: Cambridge University Press, 1990); E. Martin, *The Women in the Body* (Boston: Bacon Press, 1992) (hereafter cited and Martin, *Women)*; and A. I. Tauber, *The Immune Self: Theory or Metaphor* (Cambridge: Cambridge University Press, 1994).

12. See Martin, *Women*.

indication that the "machine" does not work anymore. This picture of "machine-failure" seems to contribute to our negative view of both processes.

Philosophers also have used metaphors to explain their ontological, epistemological, ethical, or political theories. One prevalent metaphor in ethical and political philosophy is the one defended by the liberal tradition that includes utilitarians, Kantians, and liberal individualists, namely the metaphor of egoistic, rational actors.[13] All these theories have in common the image of human beings as autonomous, self-interested individuals engaged in rational decisionmaking to develop the norms that govern social interactions and institutions. Utilitarians seem to depict people as agents trying to maximize utility, while Kantians and liberal individualists often characterize humans as actors in abstraction from any concrete historical, social, or political context.

This image of self-determining individuals who try to lead their lives largely according to their own conception of the good seems to inform one important distinction in liberal theories: the distinction between negative and liberty rights and positive or welfare rights.[14] Negative rights are rights to be free from some action that others may take. Positive rights, on the other hand, are rights that others provide a particular good or service. A negative right entails another's duty to refrain from doing something. The right to life is an example of a negative right. It requires that people refrain from killing or assaulting others. Rights to free speech and freedom of religion are also instances of negative rights. On the other hand, a person's positive rights

---

13. See, for example, T. Hobbes, *Leviathan* (Indianapolis: Bobbs-Merril, 1958); I. Kant, *The Foundations of the Metaphysics of Morals* (Indianapolis: Bobbs-Merril, 1959); J. Feinberg, *Social Philosophy* (Englewood Cliffs, NJ: Prentice-Hall, 1973); J. Rawls, *A Theory of Justice* (Cambridge: Harvard University Press, 1971) (hereafter cited as Rawls, *Justice*); R. Dworking, *Taking Rights Seriously* (Cambridge: Harvard University Press, 1977); J. S. Mill, *Utilitarianism* (Indianapolis: Hackett, 1979); D. Lyons, *Rights* (Belmont: Wadsworth, 1979); J. Locke, *Second Treatise of Government* (Indianapolis: Hackett, 1980); H. Sidgwick, *The Methods of Ethics* (Indianapolis: Hackett, 1981); C. Gauthier, *Morals by Agreement* (Oxford: Oxford University Press, 1986); J. Bentham, *An Introduction to the Principles of Morals and Legislation* (New York: Prometheus, 1988).

14. See previous note for references. See also Nancy L. Rosemblum, ed., *Liberalism and the Moral Life* (Cambridge: Harvard University Press, 1989).

entail another's duty to do something for that person. The rights to health care and education are positive ones because they stipulate that people have a duty to provide others with good and services. Although all individual-rights approaches acknowledge negative rights, not all of them admit positive or welfare rights because positive rights may place unjustifiable burdens on others.

With a conception of the individual as self-interested and self-determining, it is not surprising that the liberal individualist tradition has generally found it easier to defend negative rights over positive ones. The individual in the liberal tradition needs to be protected against the state and other people. Thus, negative rights are important because they require others to refrain from acting in ways that would violate or interfere with those rights.[15]

The metaphor of individuals as self-interested, rational actors seems also to inform the emphasis of the liberal tradition on the priority of rights over the good. Representatives of politically liberal views, such as Rawls, argue that a just society does not seek to promote any specific conception of the good, but provides a neutral framework of basic rights and liberties within which individuals can pursue their own values and life plans, consistent with a similar liberty for others.[16] Principles that do not presuppose any particular conception of the good must then govern a just society. According to proponents of the liberal tradition, such as Rawls, what justifies those principles is that they conform to the notion of right, a moral category that is prior to the good and independent of it. Liberal theorists maintain that there are two senses in which the right is prior to the good. In one sense individual rights cannot be sacrificed for the sake of welfare or the common good in order to avoid injustices (and in this aspect the liberal tradition opposes utilitarianism). In a second sense, the principles of justice that specify individuals rights cannot be premised on any particular conception of the good. They must be derived independently from the concept of right, so as to avoid imposing any specific ideal of the good on others.[17]

---

15. See prior note for references.

16. See Rawls, *Justice*.

17. See M. Sandel, "The Procedural Republic and the Unencumbered Self," *Political Theory*, 12: 1 (1984): 81-96.

This individualistic conception behind the prominence of individual rights over the common good is, however, questionable because it underestimates the basic interrelations of human beings. Individuals can prosper as moral beings and as political agents only within the context of a community. A number of philosophers have emphasized the importance of the community for the flourishing of people. Aristotle, for example, argued in the *Nicomachean Ethics* and in the *Politics* that moral and political virtue could be achieved only in the community. St. Thomas Aquinas also understood the community as the natural and necessary vehicle for meeting members' needs and gaining self-realization.[18] Also Hegel stressed the importance of different forms of community--the family, the state--for the complete realization of the political and moral capacities of human beings. The common good may, according to these philosophers, require some personal sacrifices to help and provide for others. These sacrifices, however, would work toward improving the community by, for example, providing stability, or preserving cohesion and understanding among its members.

More recently, communitarian and feminist scholars, such as Michael Sandel, Alasdair McIntyre, Annette Baier, and Elizabeth Kingdom, also have criticized the individualistic conceptions of the liberal tradition.[19] In general,

---

18. See, for example, T. Gilby, *The Political Thought of Thomas Aquinas* (Chicago: University of Chicago Press, 1958).

19. For a criticism of the liberal tradition from a communitarian point of view see, for example, M. Sandel, *Liberalism and the Limits of Justice* (Cambridge: Cambridge University Press, 1982); M. Walzer, *Spheres of Justice* (Oxford: Basic Blackwell, 1983); M. Sandel, ed., *Liberalism and its Critics* (New York: New York University Press, 1984); A. MacIntyre, *After Virtue* (Notre Dame: University of Notre Dame Press, 1984); C. Taylor, *Philosophy and the Human Sciences: Philosophical Papers* (Cambridge: Cambridge University press, 1985); D. Rasmussen, ed., *Universalism vs. Communitarianism: Contemporary Debates in Ethics* (Cambridge: MIT Press, 1990); and Shlomo Avineri and Avner de-Shalit, eds., *Communitarianism and Individualism* (Oxford: Oxford University Press, 1992). For a criticism of the liberal tradition from a feminist perspective see, for example, A. Baier, *Postures of Mind: Essays on Mind and Morals* (Minneapolis: University of Minnesota Press, 1985); S. Moller Okin, *Justice, Gender and the Family* (New York: Basic Books, 1989); C. Gilligan, *In a Different Voice: Psychological Theory and Women's Development* (Cambridge: Harvard University Press, 1989); I. M. Young, *Justice and the Politics of Difference* (Princeton: Princeton University Press, 1990); E. Kingdom, *What's Wrong with Rights: Problems for Feminist Politics of Law* (Edinburg, U.K.:

communitarians and feminists argue that the premises of individualism such as rational agents who choose freely are wrong. They claim that the only way to understand human behavior is to refer to people in their social, cultural, and historical contexts. For these scholars, human beings are, in part, products of their environments. They are elements of a state of interdependence maintained by a human network of aid, services, and restraints. Persons do not emerge fully grown and complete into the world, prepared to perform actions and form social contracts as autonomous beings. They depend on the physical and emotional effort of other persons who nurture, care, and help them. People's sense of who they are is in part a product of their social history and actual circumstances. People cannot conceive themselves as wholly detached from their communal ends and values. On the contrary, communities have a constitutive role in the formation of personal identity.

Supporters of common good approaches affirm that individualistic views of human beings give rise to morally unsatisfactory consequences. Among them are the difficulty or impossibility of achieving a genuine community, the disregard for special obligations that people hold to members of their communities--families and nations-- or the encouragement of egoistic goals.

From the perspective of humans as interrelated, caring, nurturing, and social beings, an emphasis on the good of the community comes as no surprise. If individuals are who they are in part as a result of their social relations with others, if attachment creates and sustains the human community, then a requirement for aiding and acting in ways that improve the community (as opposed to stronger obligations not to interfere with, for example, individuals' economic rights) appears justifiable.

1.2.2. INDIVIDUALS' AND COMMUNITIES' INTERESTS IN IN VITRO FERTILIZATION ASSESSMENTS

Questions about individual and social responsibilities, resource allocations, effects on women's lives, impacts on family structures, and consequences for

---

Edinburgh University Press, 1991); M. Minow, *Making All the Difference: Inclusion, Exclusion, and American Law* (Ithaca: Cornell University press, 1990); and M. A. Glendon, *Rights Talk: The Impoverishment of Political Discourse* (New York: Free Press, 1991)

societal values should all be part of an adequate assessment of IVF and related procedures. Reproduction is not an individual, isolated act with no effects on others. It is an act that involves the community. It always occurs with a partner (even if the partner is an anonymous egg or sperm donor), it creates new persons who in turn affect society, and it often requires the collaboration of medical professionals. It also may require tax payers' money, especially in countries with public health systems.

As fundamental steps in the selection of public policies about infertility treatments, IVF assessors should have evaluated the technology in ways that attended to both individual and collective interests. In doing so, they more likely would have considered a wider range of options that could help decisionmakers and the public in their choices. Contrary to this, assessors have emphasized the importance of individuals' choices and have underestimated the possible consequences of the extensive use of IVF and related technologies for the common good.

IVF evaluations seem to give priority to individuals' interests by stressing procreative rights and by neglecting any significant discussion of the social consequences of the implementation and use of IVF and related procedures. Procreative rights seem to outweigh issues of the common good in the four IVF reports. For example, the Victorian assessment stresses the importance of considering "the rights of infertile people to take advantage of IVF processes."[20] There is no relevant discussion on how the use of IVF may affect the common good. Moreover, after concluding that IVF is acceptable in the most common situation (i.e., a married couple supplying their own genetic material,) the Victorian committee affirms that some safeguards to protect the interests of the community and the individuals involved in IVF are necessary. But when discussing those safeguards, assessors mention mainly protective measures for the individuals undergoing IVF treatment such as the need for licensing IVF hospitals, counseling for the infertile couples, information and consent for those seeking treatment. There is only one measure mentioned in order to safeguard the community, namely the need

---

20. Committee to Consider the Social, Ethical, and Legal Issues Arising from In Vitro Fertilization, *Interim Report* (Victoria: Victorian Government Printer, 1982), p. 2 (hereafter cited as Victorian Report).

for educative programs to increase awareness and understanding of infertility.[21]

Similarly, the British study offers as the best argument for IVF the fact that "the technique will increase the chances for some infertile couples to have a child."[22] The report stresses the fact that "for some couples this [IVF] will be the only method by which they may have a child that is genetically entirely theirs." But assessors dismiss, without much discussion, arguments against IVF based on the creation of surplus embryos or in issues of allocation of scarce resources.[23] For instance, they consider a possible argument against IVF and related procedures based on the fact that the desire to have children is only a wish, not a need, and therefore should not be satisfied at the expense of other more urgent demands on resources. The argument is, nevertheless, discarded by saying that "there are many other treatments not designed to satisfy absolute needs which are readily available within the NHS."[24] Likewise, assessors dismiss another objection against IVF based on the use of scarce resources by arguing that although "questions about the use of scarce resources are proper questions, deserving serious consideration, essentially they relate to the extent of provision [of IVF], not to whether there should be any provision at all."[25] Despite recognizing the importance of issues of scarce resources, evaluators neglect to discuss them.

The Spanish assessors also emphasize the right to form a family.[26] With an inadequate analysis, they reject arguments against IVF based on the production of extra embryos, risks to women's health, and issues of cost.[27] Against the argument of restricting IVF and related procedures because they are too expensive to be funded by the public system, the Spanish evaluators

21. Victorian Report, pp. 21-24.

22. M. Warnock, A Question of Life. The Warnock Report on Human Fertilization and Embryology (Oxford, UK: Blackwell, 1985), p. 32 (hereafter cited as Warnock Report).

23. Warnock Report, pp. 31-32.

24. Warnock Report, p.9

25. Warnock Report, p. 32.

26. Comisión Especial de Estudio de la Fecundación *in Vitro* y la Inseminación Artificial Humanas [Special Commission for the Study of Human *in Vitro* Fertilization and Artificial Insemination], *Informe* [*Report*] (Madrid: Gabinete de Publicaciones, 1987), ch. 3 (hereafter cited as Spanish Commission).

27. Spanish Commission, pp.67-68.

object that such an argument "ignores that there are many other accepted treatments that are really expensive and benefit only a few."[28] Of course, the Spanish assessors neglect to mention those treatments and fail to address how their objection answers the question of misallocation of scarce resources. Moreover, in their recommendations, evaluators suggest that "legislation or regulations should take into account the conflicting interests of women, legal parents, donors, future children, clinics, and professional associations."[29] Assessors do not make recommendations on how to consider the interests of the community as a whole.

Finally, the United States report stresses the strong tradition of nonintrusion of the federal government into reproductive issues.[30] In fact, the assessment dedicates a whole chapter, and a section of another chapter, to the discussion of procreative rights.[31] U.S. assessors do recognize "the significance of considering the consequences of individual actions and social practices for all those affected."[32] They emphasize the fact that "any evaluation must consider the consequences of these techniques for the infertile couples, for their prospective children, and for the rest of society."[33] However, the discussion of these impacts consists of the claim that "a strong argument can be made that individuals have a duty to refrain form utilizing reproductive technologies in ways that could possibly harm future generations or make disproportionate claims on the resources of existing generations."[34] There is no, however, significant analysis of the effects of IVF and related technologies on existing or future generations, such as impacts on our conception of the family, our notions of personhood, consequences for women's or children's lives, or for the allocation of scarce resources.

In the next section I argue that IVF assessors have erred because, in seeing reproductive problems as primarily individual ones, they have underestimated the role of social, ethical, and political solutions such as

---

28. Spanish Commission, p.67.

29. Spanish Commission, p. 113.

30. Office of Technology Assessment, *Infertility: Medical and Social Choices* (Washington, D.C.: U.S. Government Printing Office, 1988), ch. 12 (hereafter cited as OTA, *Infertility*).

31. OTA, *Infertility*, ch. 12, and pp. 205-207.

32. OTA, *Infertility*, p. 212.

33. OTA, *Infertility*, p. 212.

34. OTA, *Infertility*, ch. 214.

prevention in solving fertility difficulties. Next, I defend the thesis that IVF evaluations are problematic because, in assuming too broad a definition of 'infertility', they may encourage people to use IVF when they do not need it. Undervaluing social, ethical, and political solutions to the problem of infertility, and encouraging unnecessary use of IVF therapy may have negative consequences for the common good. Assessors may promote public policies that put women at risk, that misallocate scarce resources, and that underestimate social responsibility for the community's well being.

1.3. Neglecting the Common Good: The Role of Social, Ethical, and Political Solutions to Infertility Problems

In the search for solutions to the problem of infertility, we may be acting like the drunk looking for his wallet. When asked what is he doing searching under the streetlight, the drunk answers that he is trying to find a lost wallet. Questioned whether that is the place where he lost it, he replies: "No, but this is the only place where there is light." Like the drunk, assessors of IVF and other infertility treatments may be looking in the wrong place.

In this section I shall defend the thesis that, in accepting an individualistic framework for their analyses, evaluators have erroneously defined the problem of infertility mainly as a medical problem needing a technological solution. Consequently, they have found a narrowly technical answer to it. Framing the problem of infertility mainly as an individual one is questionable for several reasons: first, because reproduction is an act that clearly involves the community and other people; and second, because the causes of reproductive difficulties and the reasons that make infertility a serious concern are, in part, socially rooted. Thus, in presenting the problem of infertility as primarily an individual one, assessors might have underestimated the role of conservative responses, such as prevention. Undervaluing conservative solutions is, however, problematic because it may foster public policies that underemphasize social influences on, and the community's responsibility for, the well being of its members and that promote unfair discrimination against women.

1.3.1. THE FALLACY OF UNFINISHED BUSINESS

Assessors often fall victim to an ethical and methodological assumption that Keniston labeled 'the fallacy of unfinished business' when they restrictedly define the problems they need to address.[35] This assumption is that technological, medical, and environmental problems have only technical, but not ethical, social, or political solutions. If assessments deal with purely technological questions then they deal solely with choosing means to a presupposed end, rather than with also evaluating alternative ends. However, if assessors take into account social, ethical, and political solutions to particular problems, then they would be able to assess different ends. For instance, if assessors, in evaluating the environmental problems caused by automobiles, analyze only the effectiveness of various pollution-control equipments, then they are investigating the means to a presupposed end, e.g., the best way to continue to use conventionally designed automobiles.[36] Evaluators fail, therefore, to analyze the impacts of alternative ends, e.g., using mass transit. As a consequence, analysts are sanctioning the use of private autos even though this mode of transport may not be a truly optimal alternative.

Because the main role of the IVF assessments is to offer guidance to decisionmakers in their tasks, commissioners should provide them with a variety of policy options. Such a variety should include social and political answers, i.e., institutional changes, as well as technological ones, i.e., infertility treatments, because often merely technical solutions do not solve the problems. In doing so, assessors would likely increase both the freedom and the power of those who are in charge of making public decisions.[37] Instead, evaluators of the four IVF reports also seem to have fallen victim to the fallacy of unfinished business. Thus, they have defined the problem of infertility in a narrow technical way and have offered mainly technological but not ethical, social, or political solutions.

---

35. See, K. Keniston, "Toward a More Human Society," in *Contemporary Moral Issues*, ed., H.K. Girvetz (Belmont, CA: Wadsworth, 1974), pp. 401-402. See, also, K. Shrader-Frechette, *Science Policy, Ethics, and Economic Methodology* (Dordrecht: Reidel, 1985), ch. 4 (hereafter cited as Shrader-Frechette, *Science*).

36. Shrader-Frechette, *Science*, ch. 4.

37. See Shrader-Frechette, *Science*, ch. 4.

IVF assessors recognize the importance of facing reproductive difficulties. They indicate the significance of the infertility problem by mentioning the prevalence of this condition in the population. All of them accept that between 8.5% and 13% of married couples are infertile.[38] Also, they refer to the relevance of the issue alluding to the suffering of infertile people. Thus, in the Victorian assessment, infertility appears as a serious, even tragic, deprivation for many of the couples affected by it.[39] The British committee presents childlessness as a source of stress. According to this report, the lack of children can be shattering for many. It can disrupt their picture of the whole of their future lives. They may feel that they will not be able to fulfill their own and other people's expectations.[40] The Spanish commissioners also suggest that infertility may be a cause of serious psychological problems that may end the unity of the couple.[41] Finally, the United States report presents the problem as a painful private experience that may also break the marital ties.[42]

Although assessors recognize infertility is a serious problem, they do not adequately analyze the issue. Hence, all four reports presuppose, without evaluation, that the problem is mainly a medical one that demands a technical solution. The Victorian report, for example, indicates that couples with reproductive difficulties have a number of medical and surgical methods available for treatment, one of which is IVF.[43] The British committee concludes that infertility is a condition meriting medical treatment.[44] For the Spanish commissioners, infertility is an illness or the consequence of an illness, with its physical, psychological, and social components.[45] Nevertheless, when analyzing possible solutions to reproductive difficulties, the Spanish assessment pays little attention to social responses, such as institutional changes or preventive measures, to reproductive difficulties.[46]

---

38. See Victorian Report, p. 4; Warnock Report, p. 8; Spanish Commission, p. 51; and OTA, *Infertility*, p.3.
39. Victorian Report, p. 4.
40. Warnock Report, p. 8.
41. Spanish Commission, p. 53.
42. OTA, *Infertility*, p. 37.
43. Victorian Report, p. 4.
44. Warnock Report, p. 9-10.
45. Spanish Commission, p. 51.
46. See Spanish Commission.

Finally, the United States report mentions that infertility is not only a personal medical problem but also in some ways a social construct. It is, assessors point out, in part a manifestation of the American commitment to a complex, pluralistic society, in which women balance childbearing, for example, with education or career goals.[47] Nevertheless, the study is limited, without reasons for it, to technologies that help establish a pregnancy.

1.3.2. UNDERESTIMATING SOCIAL, ETHICAL, AND POLITICAL SOLUTIONS TO INFERTILITY

Whether we view infertility mainly as a medical condition or also as a social one has important implications. Defining infertility as a medical difficulty suggests that a technological treatment is the appropriate response. Analyzing infertility also as a socially generated problem indicates that assessors should consider social, ethical and political solutions to reproductive difficulties. In assuming, without proper argumentation, that the problem of infertility is primarily a medical problem that requires a technological solution, assessors have focused primarily on how best to use IVF rather than on evaluating also more conservative alternatives such as prevention or institutional changes in our society.

Thus, IVF evaluators have restricted possible solutions to reproductive difficulties. Because some of the causes of the problem of infertility are socially rooted, and because reproduction is an act that clearly implicates communities and other people, presenting infertility difficulties primarily as individuals' problems is questionable. In doing so, assessors have restricted unjustifiably possible solutions such as measures that could prevent or dissolve the problem of infertility.

Preventing Infertility

One of the solutions to the infertility problem that assessors have neglected is prevention. In many cases preventive care is more effective than treatment in saving lives, reducing suffering, and raising levels of health. For example, mass vaccination programs have resulted in declines exceeding 97% in the

---

47. OTA, *Infertility*, p. 3.

incidence of mumps, measles, diphtheria, rubella, and polio.[48] Also, expenses for treatment are normally far higher than those for prevention.[49]

Many of the causes of infertility could be prevented. Sexual, contraceptive, and medical practices, occupational health hazards, environmental pollution, and food additives constitute some examples of preventable causes of infertility. Sexually transmitted diseases (STDs) such as chlamydia, gonorrhea, and syphilis are responsible for 20% of the cases of infertility.[50] Thousands of women each year have to deal with procreation problems due to pelvic inflammatory disease (PID) caused by STDs.[51] Hormonal contraceptives such as depo-provera, as well as others like intra-uterine devices, increase the risk of PID and infertility. Also, according to some professionals, iatrogenic or doctor-induced infertility is common. Problems such as infections after childbirth and postoperative infections can cause reproductive difficulties.[52] Likewise social practices such as delaying childbearing may be responsible for reproductive difficulties. Some evidence also suggests environmental pollutants and chemicals can damage the reproductive capability of both women and men. Drugs such as DES can also cause infertility.[53]

---

48. See C. E. Lewis, The Role of Prevention," in Ronald M. Andersen *et al.*, eds., *Changing the U.S. Health Care System* (San Francisco: Jossey-Bass, 1996), p. 362; (hereafter cited as Lewis, Prevention.)

49. See, for example, Beauchamp and Childress, *Principles*, ch.6; H. and L. Lindemann Nelson, Justice; and Lewis, Prevention.

50. See Ellis, G. B., "Infertility and the Role of the Federal Government," in *Beyond Baby M*, eds., D. M. Bartels, R. Priester, D. E. Vawter, and A. L. Caplan, (Clifton, NJ: Humana Press, 1990), pp. 111-130.

51. See R. Jewelewicz and E. E. Wallach, "Evaluation of the Infertile Couple," in *Reproductive Medicine and Surgery*, eds., E. E. Wallach and H. A. Zacur (St, Louis, Missouri: Mosby, 1994), pp.364 (hereafter cited as Jewelewicz and Wallach, Evaluation), p. 364; and B. A. Mueller and J. R. Daling, "The Epidemiology of Infertility," in *Controversies in Reproductive Endocrinology and Infertility*, ed., M. R. Soules (New York: Elsevier, 1989).

52. See Rowland, *Living Laboratories* (Bloomington: Indiana University Press, 1992), pp. 231, 257 (hereafter cited as Rowland, *Laboratories*).

53. See Jewelewicz and Wallach, Evaluation, p. 364; Rowland, *Laboratories*, pp. 231; and R. Koval and J. A. Scutt, "Genetic and Reproductive Engineering --All for the Infertile?" in *Baby Machine. Reproductive technology and the Commercialization of Motherhood*, ed., J. A. Scutt (Melbourne: McCulloch Publishing, 1988), pp. 33-57.

Among the poor, inadequate nutrition, poor health, and limited access to health care also contribute to reproductive problems. For example, infertility is higher in poor and minority communities[54] Black women have an infertility rate one and one-half times higher than that of white women.[55] Some of the contributing factors are a higher incidence of STDs, greater use of intrauterine devices, environmental factors (such as occupational hazards affecting reproduction), lack of access to medical treatment, nutritional deficiencies, and complications or infections following childbirth or abortion.

It is true that knowledge of the causal links of infertility is crucial in preventive medicine. It is also the case that such knowledge was, and still is, limited. However, some important information has existed, about sexually transmitted diseases, contraceptive practices, and environmental hazards that could have been used by the committees in their assessment of IVF. Only the United States report gives any attention to information related to the causes of infertility and its prevention.[56] Likewise, where evidence about the causes of infertility was lacking, those reports with the task of recommending could have advised researchers to gather information so that, analysts could better evaluate the role of prevention. Only the Victorian report recommends that the government institute a campaign of public education, on the nature, causes, prevention, and treatment of infertility.[57] Nevertheless, it does not suggest any specific preventive measures or funding.

Had assessors framed the problem of infertility not only as an individual one but also as a social issue, they might have paid more attention to preventive measures rather than only to curative treatments. In focusing on the individual problem, assessors have emphasized the right to choose available technologies. However, had IVF evaluators presented infertility also as a social issue they could, for instance, have recommended stricter controls for environmental pollutants and chemicals, more research funding for safer

---

54. See M. S. Henifin, "New Reproductive Technologies: Equity and Access to Reproductive Health," *Journal of Social Issues* 49:2 (1993): 61-74 (hereafter cited as Henifin, Equity); and OTA, *Infertility*.

55. See, for example, OTA, *Infertility*, p. 51. See, also, L. Nsiah-Jefferson, "Reproductive Laws, Women of Color, and Low-Income Women," in *Reproductive Laws for the 1990s*, eds., S. Cohen and N. Taub (Clifton, NJ: Humana Press, 1989), pp.23-67 (hereafter cited as Nsiah-Jefferson, Reproductive Laws).

56. OTA, *Infertility*, ch. 4-5.

57. Victorian Report, p. 27.

contraceptives, and educational programs to prevent STDs and for treating them before they cause reproductive difficulties. Assessors also could have evaluated the costs and benefits of proposals such as an increase in funding for child care, better access to it, parental leave from employment, or job-sharing. If women know that they have access to adequate and affordable child care, or that they will not be penalized in their jobs for maternal leave, fewer women would have to postpone their pregnancy decisions as a consequence of, for instance, pursuing a career. Evaluators also could have recommended strategies to encourage men to assume more responsibilities for child and home care.

Dissolving the Problem

There are also social factors that make involuntary childlessness a serious problem.[58] Thus, changing those factors could also have a positive effect on the problem of infertility by dissolving it, or making it less onerous. Some of these social elements are pronatalistic pressures on women to reproduce, the strong emphasis of our culture on having genetically related children, and the inextricable ties between womanhood and motherhood.

Had assessors taken social considerations into account in their analyses, they could have recommended measures directed to offer different social viewpoints and practices that would make infertility appear as less problematic. Modifying the view of motherhood as the primary role of women, understanding maternity as a possible but not as a necessary choice would likely decrease the pressure on infertile women. Likewise, facilitating

---

58. See, for example, R. Arditti, R. D. Klein, and S. Minden, *Test-Tube Women. What Future for Motherhood* (London: Pandora Press, 1984); C. Overall, *Ethics and Human Reproduction. A Feminist Analysis* (Boston: Allen & Unwin, 1987), ch. 7; M. A. Warren, "IVF and Women's Interests: An Analysis of Feminist Concerns," *Bioethics*, 2: 1 (1988): 37-57; R. Achilles, "Desperately Seeking Babies: New Technologies of Hope and Despair," in *Delivering Motherhood*, eds., K. Arnup, *et al.* (London: Routledge, 1990), pp. 284-312; B. K. Rothman, *Recreating Motherhood. Ideology and Technology in a Patriarchal Society* (New York: W.W. Norton & Company, 1990); A. Phoenix , A. Woollett, and E. Lloyd eds., *Motherhood. Meanings, Practices, and Ideologies* (London: Sage, 1991); F. Laborie, "Social Alternatives to Infertility," in *Tough Choices*, eds., P. Stephenson and M. G. Wagner, pp.37-49; M. S. Ireland, *Reconceiving Women* (New York: The Guilford Press, 1993); and R. Jackson, *Mothers Who Leave* (London: Pandora, 1994).

adoption or favoring different forms of mothering may have repercussions on the infertility problem. If adoption were an accessible alternative, many couples could decide to adopt instead of undergoing technological treatment. The strong emphasis on genetic relationships, and the long-waiting periods may preclude many people from adopting children. Negative attitudes against particular races may influence decisions about adoption. For example, although some argue that the number of available children is decreasing, they have in mind the number of healthy white babies. Nonwhite children and children with various disabilities are, however, available. Focusing on other forms of mothering also may have influences on the problem of infertility. Taking care of others' children could satisfy the desires of many for having a child. Seeing children as a good for whom all of us need to be concerned, rather than viewing then as private property of their parents, could make reproductive difficulties appear as less problematic.

1.3.3. CONSEQUENCES OF UNDERVALUING SOCIAL, ETHICAL, AND POLITICAL SOLUTIONS

In framing the problem of infertility as an individual one and focusing mainly on nonconservative solutions such as IVF, assessors have erred because they may have encouraged public policies that could harm the community. First, they may have persuaded governments to allocate resources mainly to IVF rather than to preventive medicine as solutions to infertility. Public policies that discourage preventive measures are problematic because they disregard social influences on, and communities' responsibilities for, the well being of their members. Second, evaluators may have discouraged funding for researching the causes of infertility. Such policies are questionable because they may promote unnecessary use of IVF and other infertility treatments and may foster discrimination against women.

If evaluators present IVF as the best way to solve reproductive difficulties, and if they neglect the importance of more conservative solutions such as preventive measures or institutional changes, then governments likely would fund IVF rather than prevention of infertility. Such an action would be questionable both for economic and for ethical reasons. Economically, prevention of reproductive problems arguably would be more cost effective, at least on a long-term period, than curative medicine. Investment in prevention would solve the problems of many more infertile people because the IVF success rates are very low (around 10%) and the treatment is very expensive. The average cost of a child born by IVF is around $66,500, excluding the costs of previous treatments and the post-natal expenses.[59] According the World Health Organization, the costs of one IVF treatment could prevent infertility in 1000 women.[60] Funding preventive measures could include, as I have mentioned, educative programs about the causes of infertility, screening for STDs, and funding research for safer contraceptives.

There are also compelling ethical reasons to foster conservative solutions such as prevention. Because, as we have seen, the sources of infertility and the reasons that make reproductive difficulties a serious adversity may be, in part, socially rooted, public policies that promote preventive measures would be more reasonable than those that focus only on curative solutions. Policies that encourage prevention likely would recognize social influences on, and social responsibility for, the well being of the members of the community. For example, the community is responsible for threats produced by environmental and occupational hazards, the utilization of chemicals and drugs, and the use of food additives that may cause infertility. To that extent, the community should be responsible for addressing reproductive problems.

Public policies that foster mainly nonconservative solutions such as IVF, rather than also conservative ones such as prevention, are questionable because prevention may protect people against unnecessary suffering while IVF cannot. Thus, because reproductive difficulties produce serious emotional costs, preventive measures would reduce affliction and raise levels of health. Furthermore, some infertility treatments are invasive and may entail significant risks to women's health. Hence, forms of health care that give preference to prevention of infertility should receive priority in the allocation of health care resources.

Moreover, nonconservative solutions such as IVF are more likely to exacerbate patterns of injustice among rich and poor. Thus, although infertility is higher in groups such as poor and minority communities, people

---

59. P. J. Newman, S. D. Gharib, and M. C. Weinstein, "The Cost of a Successful Delivery with *in Vitro* Fertilization," *The New England Journal of Medicine* 331 (July 1994): 239-243.

60. World Health Organization, *Recent Advances in Medically Assisted Conception: Report of a WHO Scientific Group* (Geneva: WHO, 1992).

who can use IVF and related procedures are mostly white couples from the upper-middle class.[61] Preventive measures, on the contrary, would more likely favor also those in underprivileged positions.

Allocating funds for prevention may seem less reasonable than financing curative treatments because the latter meets the needs of identifiable individuals. Preventive programs also help people, but it is difficult to identify them. People sympathize with someone anxiously awaiting a medical intervention such as IVF to have a child. It is more difficult to generate empathy for the unidentifiable individual whom prevention programs may help, even if the suffering avoided is equally great. Nevertheless, relocating funds for prevention does not mean that infertility treatments such as IVF would not be available anymore. As I have said before, I am not suggesting that governments should stop the use of these kinds of procedures. My argument is that, had assessors considered other options, they could have encouraged programs that could prevent harm to a great number of people.

Another consequence of IVF assessors' emphasis on technological solutions and their disregard for preventive measures is that they may have discouraged funding for researching the causes of reproductive problems. People seek infertility treatments such as IVF and related procedures after they find they have difficulties conceiving a child. Funding curative treatments mainly requires recognizing that problems of infertility exist. However, financing preventive measures requires knowledge of the causes of infertility. Therefore, undervaluing the role of prevention may have a negative effect on funding for infertility investigations.

Lack of financing for research on the causes of infertility is problematic because it may promote unnecessary use of IVF and related procedures. Even today, the occurrence of unexplained infertility may range from 6% to 60%.[62] This means that following a thorough evaluation, doctors cannot find in these couples an identifiable cause for their problems. If governments devoted more money to investigate the causes of infertility, more likely the rate of unexplained infertility would decrease. Thus, many women might not have to undergo unnecessary and risky infertility treatments because other less risky and less expensive answers may be available.

---

61. See Henifin, Equity; and OTA, *Infertility*.
62. See Jewelewicz and Wallach, Evaluation.

Discouraging funding for researching the causes of infertility is also problematic because assessors might have fostered sexist practices. Thus, although in about half of the couples with reproductive problems there is a contributing male factor (i.e., low-sperm count or poor semen quality), there is still deficient knowledge of the causes of male infertility.[63] Except in cases of the grossest anomalies, such as lack of sperm or of sperm motility, scientists do not have a clear idea about the significant factors in the ability of a man to help to produce a pregnancy. Consequently, medicine is unable to address male infertility in most cases. As a consequence of the ignorance of male reproductive problems, doctors use IVF and other new assisted-conception techniques exclusively in women. Instead of trying to overcome male problems with better and more research, professionals have routinized doctoring women to the degree that treating them for diverse male conditions is now acceptable and habitual. In such cases, a perfectly healthy, fertile woman will undergo long and hazardous treatments as a way to overcome male-related disorders.[64] Infertility is the only disorder that requires doctors to treat a healthy person --a woman-- in order to solve the problem in another one --her husband or partner. Thus, in neglecting the role of prevention, assessors may influence governments to choose policies, (i.e.,

---

63. See, for example, D. M. Colin, "Clinical Male Infertility. The Choice of Approaches for Pregnancy," *Reproduction, Fertility, and Development*, 6:1 (1994): 13-18; G. R. Cunningham, G.R., "Male Factor Infertility," in *Reproductive Medicine and Surgery*, eds., Wallach and Zacur, pp. 399-414; R. D. Kempers, "Where Are We Going?," *Fertility and Sterility*, 62:10 (1994): 686-689; J. S. Sherman, "A Modern View of Male Infertility," *Reproduction, Fertility, and Development*, 6:1 (1994): 93-104; N. E. Skakkebaek, A. Giwercman, and D. de Kretser, "Pathogenesis and Management of Male Infertility," *The Lancet*, 343:8911 (1994): 1473-1478; and J. D. McConnell, "Diagnosis and Treatment of Male Infertility," in *Textbook of Reproductive Medicine*, eds., B. R. Carr and R. E. Blackwell (Norwalk, Connecticut: Appleton & Lange, 1993), pp. 453-468.

64. See, for example, M. Sigman, "Assisted Reproductive Techniques and Male Infertility," *The Urologic Clinic of North America*, 21:3 (1994): 505-515; D. Royere, "Assisted Procreation for Male Indication," *Rev. Prat.*, 43:8 (1993): 981-986; S. Gordts *et al.*, "Role of Assisted Fertilization Techniques in the Management of Male Infertility," *Contracep Fertil Sex* 21:10 (1993): 695-700; and R. B. Meacham and L. I. Lipshultz, "Assisted Reproductive Technologies for Male Factor Infertility," *Current Opinion in Obstetrics and Gynecology*, 3:5 (1991), pp. 656-661.

allocating resources mainly on curative treatments such as IVF) that might promote sexist practices.

As we have seen, in assuming that infertility problems are best solved through nonconservative technologies such as IVF, assessors might have fostered the selection of wrong public policies such as investing money primarily in curative treatments rather than also in preventive measures. Such policies may be questionable because they disregard communal influences on, and responsibilities for, people's well being, and because they may discourage funding research on the causes of infertility. Discouraging financial support for infertility research is problematic because it might promote unnecessary use of IVF and might foster discriminatory practices against women.

1.4. Neglecting the Common Good: The Definition of 'Infertility'

A second problem with IVF assessments is related to the assumption that the standard definition of 'infertility' adequately represents the infertile population. I argue here that assessors of IVF have erred because, in assuming too broad a definition of 'infertility', they might persuade more people to use this procedure when they do not need it. Thus, assessors seem to have conceded more importance to individuals' desires for having children than to problems of the common good. However, unnecessary use of IVF may affect the common good because it may put women at needless risk and it may encourage misallocation of public health expenditures. Because I have dealt with ethical problems of fostering policies that may put women at unnecessary risk in chapter three, I shall concentrate here on problems related to funding for IVF treatment.

1.4.1. DEFINING 'INFERTILITY'

The choice of the definition for a disease may have critical consequences, not only clinical but also social and ethical. In 1987, the Centers for Disease Control (CDC) in the U.S. developed a case definition for AIDS.[65] The CDC

---

65. See, OTA, *The CDC's Case Definition of AIDS: Implication of the Proposed Revisions* (Washington, D.C.: U.S. Government Printing Office, 1992) (hereafter cited as OTA, *Definition).*

created this definition only as a surveillance tool for determining the scope of the AIDS epidemic. Physicians, nevertheless, have used the CDC's case definition of "AIDS" as a clinical definition. Moreover, it also has been employed in research protocols, in the allocation of federal funds, and as a measure of disability in benefit programs.

The 1987 CDC's definition of "AIDS" received several criticisms because it did not include severe manifestation of the human immunodeficiency virus (HIV) that researchers have found in women and intravenous drug users; as a result, researchers underestimated the impact of the epidemic in these populations. This exclusion is of special concern because a disproportionate number of HIV-infected women and intravenous drug users are African American or Hispanics.[66]

As with the AIDS case, the choice of a definition of 'infertility' also may have not only clinical, but also social and ethical impacts. If the definition is too strict, a number of people may be excluded from it. This would prevent them from seeking an early treatment. If, on the other hand, the definition is too broad, it may result in overdiagnosis and many people could undergo risky treatment unnecessarily.[67] Therefore, knowledge about the prevalence of infertility is relevant for appropriating funds for research, medical services, and prevention programs.

According to the standard medical definition of 'infertility', a person is infertile if she or he is unable to conceive after one year of regular intercourse without contraception. Experts, however, usually do not specify what 'regular' means (once a day, once a week, etc.). Assessors have assumed, without appropriate data, that this definition adequately represents the infertile population. Presupposing such a definition, the Victorian report mentions the fact that in Australia infertility affects 10% of married couples.[68] The British report cites the commonly quoted figure according to which one couple in ten is childless.[69] The Spanish Commission accepts the fact that in

66. See, OTA, *Definition*. The revision in 1991 of the 1987 case definition of AIDS has not been less problematic, however.

67. I will assume here, as the four reports analyzed do, that infertility is a condition in need of medical treatment.

68. Victorian Report, p.4.

69. Warnock Report, p.8.

Spain between 10% and 13% of all married couples are sterile.[70] Finally, the United States assessment refers to the estimated 8.5% of married couples in the U.S. that are infertile.[71] It is true that the British and the United States reports recognize the lack of accurate statistics on the prevalence of infertility.[72] They mention the difficulty of obtaining information about reproductive problems and the limitation of the available survey data. Nevertheless, they draw their conclusions based on statistics that may misrepresent the number of infertile people.

The identification of infertility is not, however, as straightforward as it may seem. A continuum ranging from sterility through subfertility and infecundity to fertility characterizes the ability to conceive. In the statistical data on infertility, experts integrate data on sterility (incapacity to procreate) with those on infertility (difficulty in becoming pregnant within one or two years), and infecundity (inability to achieve a live birth). Because of the arbitrariness of the terms, doctors diagnose as infertile a substantial proportion of people that, without treatment, would be able to conceive a child. For example, in using the definition of one year of unprotected intercourse (with no specification of the frequency of the sexual act), 79% to 84% of couples meeting such a definition will eventually, without treatment, have children.[73] Also, according to some recent studies, the rate of pregnancy of untreated couples diagnosed as "infertile" is only 6% less than that of the treated infertile couples. Some studies suggest that the proportion of pregnancies that seem to be independent of treatment ranges from 21% to 62%.[74] Furthermore, evidence shows that 40% of couples with untreated unexplained infertility ultimately conceive children within three years. Likewise, according to other analyses, 64% of women with primary unexplained infertility, and 79% with secondary infertility conceive a child within nine years.[75]

---

70. Spanish Commission, p.51.

71. OTA, *Infertility*, p.3.

72. See Warnock Report, p.8, 13; and OTA, *Infertility*, p.35.

73. See OTA, *Infertility*, p. 35.

74. See, HealthFacts, *Infertility Treatments: A Demand for More Honesty*, 19 (1994): 1-4. See also, F. Laborie, "Social Alternatives to Infertility," in *Tough Choices*, eds., Stephenson and Wagner, pp. 37-61.

75. See Jewelewicz and Wallach, Evaluation, pp.364;

1.4.2. CONSEQUENCES OF ASSUMING TOO BROAD A DEFINITION OF 'INFERTILITY'

If the standard definition of 'infertility' results in overdiagnosis of people with reproductive difficulties, then it may lead to inflated estimates of the prevalence of reproductive problems in the community. In accepting such estimates, assessors may have promoted needless use of IVF. As a consequence they may have encouraged policies that promote using health resources in ways that do not serve the common good as well as they might.

No doubt, infertility is a stressful and painful experience for many people. Most human beings value highly the fact of having children. If to this we add that a significant number of the population suffers from reproductive difficulties, then it is arguable that most democratic countries would devote health care resources to try to solve these problems. In assuming a too broad definition of infertility, assessors of IVF likely have erroneously legitimized excessive funding for infertility treatments. If health-care resources were not scarce, there would be no problem. However, this is not the case. Therefore, when governments decide to invest money in infertility treatments, they lose the chance to spend those same funds on other health-care measures. Thus, IVF might have great opportunity costs. Hence, excessive financing of IVF and related procedures is questionable because, given that the definition of 'infertility' results in overdiagnosis, many women may use treatments that they do not need, and therefore governments might spend money inefficiently. Such a policy would likely harm the common good.

Of course, the use of unnecessary technologies is not an exclusive problem of IVF and related techniques. Utilization of other medical procedures appears also as unnecessary.[76] For example, with a 24% rate of performance, caesarian section is the leading major surgical procedure performed in U.S. hospitals.[77] However, the known and agreed obstetric

---

76. See, for example, Consumer Reports.

77. See, for example, R. S. Stafford, "Alternative Strategies for Controlling Rising Cesarean Section Rates," *JAMA,* 263: 5 (Feb. 1990): 683-88; Consumer Reports, p. 440; C. Sakala, "Medically Unnecessary Caesarian Section Births: Introduction to a Symposium," *Social Science and Medicine*, 37: 10 (Nov. 1993): 1177-98 (hereafter cited as Sakala, Unnecessary Caesarian); A. K. LoCicero, "Explaining Excessive Rates of Cesarean and Other Childbirth Interventions: Contributions from Contemporary Theories of gender and Psychosocial

indications for this operation suggest that a rate of 6-8% would be adequate.[78] Average hospital charges and physician fees for a caesarian birth exceed those of a vaginal birth by around $3,000. Thus, if the rate of caesarian section had been 6% instead of 24%, over 731,000 caesarians and $2.1 billions in expenditures would have been eliminated.[79] Many other nations, such as Canada or Australia, have exhibited a similar trend of a sharply rising caesarian-section rate in recent decades.[80] According to several studies, between 40% and 80% of the caesarian sections that doctors perform, afford no benefit to either the baby or the mother.[81] Furthermore, unnecessary caesarian sections put mothers and babies at risk of medical and psychological complications.[82] With little exception, women who are healthier, of higher social class, better insured, and/or cared for in private facilities are at considerably higher risk for caesarian birth than their counterparts who are less healthy, of lower socioeconomic status, under- or uninsured, and/or cared for in public facilities.[83]

Development," *Social Science and Medicine*, 37: 10 (Nov. 1993): 1261-69.

78. See, for example, C. Francome and W. Savage, "Caesarian Section in Britain and The United States 12% or 24%: Is Either the Right Rate?" *Social Science and Medicine*, 37: 10 (Nov. 1993): 1199-218.

79. See, for example, Sakala, Unnecessary Caesarian, p. 1183.

80. See, for example, F.C. Notzon, "International Differences in the Use of Obstetric Interventions," *JAMA*, 263, (1990): 3286-91; Sakala, Unnecessary Caesarian, pp. 1186-93.

81. See, for example, U.S. Department of Health and Human Services, Public Health Service, National Institutes of Health, *Caesarian Childbirth* (Bethesda, Maryland: NIH, 1981); C. Jones, *Birth Without Surgery* (New York: Dodd, Mead & Co., 1987); American Public Health Association, "Reduction of Unnecessary Caesarian Section Births," *American Journal of Public Health*, 80 (1990): 225-27; Consumer Reports, p. 440.

82. See, for example, C. Mutryn, "Psychosocial Impacts of Caesarian Section on the Family: A Literary Review," *Social Science and Medicine*, 37: 10 (Nov. 1993): 1271-81; and E. Shearer, "Caesarian Section: Medical Costs and Benefits," *Social Science and Medicine*, 37: 10 (Nov. 1993): 1223-31.

83. See, for example, J. B. Gould and R. S. Stafford, "Socioeconomic Differences in Rates of Cesarean Section," *The New England Journal of Medicine*, 321 (1989): 233-39; R. S. Stafford, "The Impact of Nonclinical Factors on Repeat Cesarean Section," *JAMA*, 265 (1991): 59-63; C. Sakala, Unnecessary Caesarian, p.1180.

Another procedure that doctors perform unnecessarily in many cases is hysterectomy.[84] This is the second most common major surgical procedure in the U.S.. According to some studies, between 24% and 30% of hysterectomies are unnecessary. Many gynecologists routinely recommend this procedure for fibroid (benign growths on the wall of the uterus), uterine prolapse (a portion of the uterus descend through the vaginal opening), and heavy bleeding. However, less drastic and safer alternative treatments are available for all three conditions. Furthermore, up to one half of all women undergoing hysterectomy experience complications, and up to 15% will need a second surgery for complications related to the original operation. Researchers estimate that, in the absence of cancer or pregnancy, death rates immediately after hysterectomy range from 6 to 11 per 10,000. In a series of studies, the incidence of nonfatal complications of this procedure ranges from 25% to 50% of cases. Some of the complications are hemorrhage requiring transfusion, injury of adjacent organs, and pulmonary embolism (blood clot to the lungs). Some of the long-term effects of hysterectomy are premature ovarian failure, persistent pelvic pain, urinary symptoms, and depression. There are also significant financial implications of the overuse of this procedure. Thus, in 1991, according to some studies, the average hospital cost for a hysterectomy was approximately $7,600. Such costs do not include physicians' fees.[85]

As with the cases of caesarian sections and hysterectomies, unnecessary use of fertility treatments may result from an inadequate diagnosis of reproductive difficulties. Such unnecessary use is problematic because it may increase excessive health-care expenditures. For example, according to some studies, for couples with a better chance of successful delivery with IVF, i.e., those with a diagnosis of tubal disease, the procedure costs $50,000 per delivery for the first cycle and $72,000 for the sixth (excluding previous infertility treatments, obstetric and perinatal costs). However, for couples in

---

84. See, for example, Consumer Reports, p. 440; Committee on Labor and Human Resources, *Unnecessary Hysterectomies, The Second Most Common Major surgery in the United States* (Washington, D.C.: U.S. Government Printing Office, 1993) (hereafter cited as CLHR, Hysterectomies); and C. C. Nadelson and M. T. Notman, "Women: Health-Care Issues," in *Encyclopedia of Bioethics*, ed. W. T. Reich, (New York: Simon & Schuster Macmillan, 1995), vol. 5, pp. 2560-2572.

85. See, for example, CLHR, Hysterectomies.

which the woman is older and there is a diagnosis of male-factor infertility, the cost of IVF rises from $160,000 for the first cycle to $800,000 for the sixth.[86] Nevertheless, studies suggest that non-IVF therapy is still worthwhile in cases of male, cervical, and ovulatory factors, even in patients with more than three years duration of infertility. Alternatives such as artificial insemination are less expensive than IVF but more or equally effective.[87]

In countries like the U.S., with a large private-health sector, receipt of IVF and many other infertility services depends mainly on the individual's ability to pay. Nevertheless, a small number of IVF clinics treat many Medicaid patients. Although the majority of U.S. health- insurance plans have specifically excluded coverage of IVF from their policies, there appears to be a significant amount of reimbursement for the various components of IVF treatment, such as ovulation-induction drugs or laparoscopy. The same seems to be the case for U.S. government reimbursement's plans.[88] Thus, in the United States private and public infertility expenditures in 1987 totaled about $1.0 billion. Public and private disbursement for IVF amounted to $66 million.[89] In 1994, costs of infertility services were over $2 billion.[90] Around 10% of U.S. infertility expenditures come from public resources.[91]

In countries like Spain, Britain, and Australia, where governments' subsidies cover part or all the costs of IVF, all taxpayers bear the burden of paying for an expensive service that benefits only a few. For example, in Australia total costs of IVF alone in 1987 were $30 million. The direct annual costs to the Australian government were $17 million. In 1991, the annual costs of IVF and related services were $35 million, and the government was paying 75% of these expenses.[92]

---

86. See P. J. Newman, *et al.*, "The Costs of a Successful Delivery with In Vitro Fertilization," *The New England Journal Of Medicine*, 331:4 (1994): 239-43.

87. Royal Commission on New Reproductive Technologies, *Proceed with Care* (Ottawa, Canada: Canada Communications Group, 1993) (hereafter cited as RCNRT, *Care*) .

88. OTA, *Infertility*, p. 161.

89. OTA, *Infertility*, p. 161.

90. S. Brownlee, "The Baby Chase: Millions of Couples Have Infertility Problems, and Many Try High-Tech Remedies. But Who Minds the Price Clinics They Turn to?," *News and World Report* 117: 22 (1994): .84.

91. OTA, *Infertility*, p. 147.

92. See, D. Bartels, "The Financial Costs of In Vitro Fertilization: An Example

Even if IVF were fully privately funded, its use could still raise the expenditures in the public system because of the indirect medical costs resulting from IVF treatment. First, there may be complications due to the procedures and drugs used for IVF. Treatment of such complications may result in costs for the public system. For example, in cases of infections, excessive bleeding, spontaneous abortions, ectopic pregnancies, or hyperstimulation of the ovaries, the woman could go to the public system to look for care. Second, because of the frequency of multiple births when using this technique, there are more caesarian sections, more premature deliveries, and more low-birth weight children, all resulting in increased health care and social costs. Besides this, there are also expenses associated with many of the children conceived through IVF. The rate of IVF children with low birth weight is eleven times higher than in the general population. Preterm birth and low birth weight remain important causes of infant death, long-term physical disability, mental retardation, and hearing and vision impairment.[93] Studies show that multiple-gestation pregnancies, a high proportion of which results from IVF, dramatically increase hospital charges. According to one of these studies, if all the multiple gestations resulting from assisted-reproduction techniques had been singleton pregnancies, the predicted saving to the health care system in the examined hospital alone would have been over $3 million per year.[94] In Australia, the costs to the government for

---

from Australia," in *Tough Choices*, eds., Stephenson and Wagner, pp.73-82 (hereafter cited as Bartels, Costs); and Rowland, *Laboratories*, p. 41.

93. See, for example, P. Stephenson, "Ovulation Induction During Treatment of Infertility: An Assessment of the Risks," in *Tough Choices*, eds., Stephenson and Wagner, pp. 97-121; L. Koch, "Physiological and Psychological Risks of the New Reproductive Technologies," in *Tough Choices*, eds., Stephenson and Wagner, pp. 122-134; J-P. Relier, M. Couchard, and C. Huon, " The Neonatologist's Experience of In vitro Fertilization Risks," in *Tough Choices*, eds., Stephenson and Wagner, pp. 135-143 (hereafter cited as Relier, Couchard, and Huon, Neonatologist); J. Jarrel, J. Seidel, and P. Bigelow, "Adverse Health Effects of Drugs Used for Ovulation Induction," in *New Reproductive Technologies and the Health Care System. The Case for Evidence-Based Medicine*, Royal Commission on New Reproductive Technologies (Ottawa, Canada: Canada Communications Group, 1993), pp. 453-549.

94. See, T. L. Callahan *et al.*, "The Economic Impact of Multiple-Gestation Pregnancies and the Contribution of Assisted-Reproductive Techniques to Their Incidence," The *New England Journal Of Medicine*, 331:4 (1994): 244-49. See also Relier, Couchard, and Huon, Neonatologist.

neonatal intensive care of babies born through IVF and related procedures were A$4.1 million in 1986.[95]

Moreover, even in countries such as the United States, where public money for IVF and associated techniques is limited, societal interdependencies and professional contracts have created and enhanced doctors' abilities to use IVF. They employ tools and technologies developed in part through societal resources. Also, public money supports physicians through learning, because virtually no student, even in private schools, pays for the full costs of education; taxes or donations usually supplement that cost.[96]

As I have shown, in assuming that the standard definition of 'infertility' adequately represents the population with reproductive problems, assessors may have encouraged more people to use this technology when they do not need it. As a consequence, they may promote public policies that may harm the common good, because governments could increase unnecessarily the amount of resources dedicated to this technique. Promoting excessive funding of IVF is questionable because IVF and related technologies, as we have seen, are expensive procedures. If we assume, I think safely, that health resources are scarce, then allocating money for IVF might prevent funding other health-care measures that are equally or more important, such us prenatal care, infertility prevention, or mothers' and children's health care.

1.5. Some Objections and Responses

In response to my analysis of IVF assessments, critics may pose several objections. First, they can argue that showing that social, ethical, and political solutions (such as prevention) are cost-effective is extremely arduous because of the difficulty of obtaining hard data about them. Second, they can object that these kinds of answers to reproductive problems may require unattainable institutional changes. Third, critics may argue that denouncing IVF assessors for assuming too broad a definition of 'infertility' is incorrect

95. See Bartels, Costs, p. 80.

96. See, K. Shrader-Frechette, *Ethics of Scientific Research* (Lanha, Md.: Rowman and Littlefield Publishers, 1994), ch.4 (hereafter cited as Shrader-Frechette, *Science*).

because it ignores the fact that, given the importance that people place on having children, analysts have used a definition that more likely would enhance the welfare of a higher number of individuals. Fourth, critics might argue that because analysts require some data in order to arrive at their conclusions, and given the difficulties of realizing new investigations on the prevalence of infertility, it is reasonable for them to use the available information. Fifth, they may say that assessors have adopted such a definition because it is the standard medical definition and most people seeking infertility treatment also share it. I shall treat these objections in order.

1.5.1. DIFFICULTY OF OBTAINING HARD DATA

Although the first objection (that showing that social, ethical, and political solutions such as prevention are cost-effective is extremely arduous because of the difficulty of obtaining hard data about them) seems particularly compelling, it fails for several reasons. First, the difficulty of obtaining hard data on untested technologies and policies is a conspicuous problem with any new option. Thus, this criticism ignores the fact that if one dismisses novel alternatives because of the difficulty of obtaining hard data, then researchers would unlikely generate more data. Therefore, analysts would never study new policies and procedures, and technological and social progress would be arduous, if not impossible. Obviously, this result is undesirable.

Furthermore, it is true that there is incomplete evidence about alternatives to the infertility problem, such as prevention, making these programs difficult to assess. For example, many prevention programs look for alterations in individuals' behavior, and both the changes and whether the modifications in people's comportment produced the intended results are difficult phenomena to measure. Likewise, many different factors such as peer pressure, home environment, and media images influence people, making it problematic to assess whether the prevention program or these other factors were responsible for the change of behavior. However, many medical treatments are equally complicated to analyze. For example, coronary bypass surgery in patients over 65 years of age is a medical treatment difficult to evaluate because other factors such as social support, diet, exercise, tobacco use, or socioeconomic status influence the health of patients who have undergone this procedure.[97] Thus, even over a short period of time, evaluating reduced incidences of particular conditions (such as sexually transmitted diseases) in a population that has received a prevention program is not necessarily more difficult than assessing medical treatments.

Also, the difficulty of obtaining hard data may be resolved, in part, using information about other preventive practices.[98] For instance, immunization is one of the greatest triumphs of science in the prevention of diseases in childhood. Mass vaccination programs have resulted, as I have said, in declines over 97% in the incidence of mumps, measles, polio, diphtheria, and rubella.[99] The attempt to prevent cardiopulmonary diseases attributable to cigarette smoking has generated significant findings as well.[100] Similarly, investigations show that AIDS education that combines skills building and information and motivates people to use these skills has reduced significantly high-risk behavior. There also is evidence to show that some programs are effective in preventing exposure to some of the risk factors associated with infertility such as smoking. For instance, a comprehensive strategy to decrease smoking among Canadians, including legislation such as banning it in public places, school programs, public education, and high taxation of tobacco, has significantly reduced the incidence of this risk factor.[101]

---

97. See, RCNRT, *Care*, ch. 15.

98. See, for example, U.S. Preventive Task Force, *A Guide to Clinical Preventive Services: An Assessment of the Effectiveness of 169 Interventions* (Baltimore, Md.: Williams and Wilkins, 1989); U.S. Department of Health and Human Services, Public Health Service, Office of Disease Prevention and Health Promotion, *Clinician's Handbook of Preventive Services* (Washington, D.C.: U.S. Government Printing Office, 1994).

99. See, for example, Lewis, Prevention.

100. See, for example, Centers for Disease Control and Prevention (CDC), "Cigarette Smoking Among Adults --United States 1991," *Morbidity and Mortality Weekly Reports* 42 (April 12, 1993):230-233; CDC, "Smoking Control Among Health Care Workers: World No-Tobacco Day," *Morbidity and Mortality Weekly Reports* 42 (May 21, 1993):365-367; and CDC, "Cigarette Smoking Among Adults in the United States --1992,and Changes in the Definition for Current Cigarette Smoking" *Morbidity and Mortality Weekly Reports* 43 (May 20, 1994):42-47;

101. See, RCNRT, *Care*, ch. 15; and Canada, Statistics Canada, *Production and Disposition of Tobacco Products* (Ottawa: Minister of Supply and Services Canada, 1989).

Similarly, arguing that we cannot implement social, ethical, and political solutions such as prevention (because of the difficulty of obtaining hard data to show that they are cost-effective) is inconsistent with the current use of many techniques and drugs for which evidence either does not exit or is difficult to assess. The cases of IVF and related procedures are some of the examples of medical techniques whose evaluations are problematic. Requiring a greater burden of proof for prevention programs seems, therefore, unjustifiable.[102]

1.5.2. UNATTAINABLE INSTITUTIONAL CHANGES

The second objection (that social, ethical, and political solutions to the problem of infertility would require unattainable institutional changes,) focuses on important practical points. Some alternatives to the problem of infertility would likely require social transformations such as changes in attitudes toward women and motherhood or alterations in family structures. Prevention would also involve educative programs, social services, legislative changes in occupational health and safety, and environmental legislation. These changes are difficult because of economic pressures and the extensive time period required to obtain results.

Although this criticism raises important concerns, it is incorrect for several reasons. First, if analysts discard policy options before they adequately evaluate them, it is difficult to see how assessors can talk about "unattainable" institutional changes. To affirm that alternatives to the problem of infertility such as prevention require infeasible changes without giving some evidence for such argument is then problematic. Moreover, if assessors only take into account policy options that are highly feasible under the technological status quo, then their evaluations erroneously encourage a self-fulfilling prophecy sanctioning current conditions, regardless of their worth.[103]

---

102. See, for example, RCNRT, *Care*, ch. 15.

103. See, Shrader-Frechette, *Science*, ch. 4.

1.5.3. IMPORTANCE OF HAVING CHILDREN

A third objection to my analysis of IVF assessments is that denouncing IVF assessors, for assuming too broad a definition of 'infertility', is incorrect. According to this objection, my criticism ignores the fact that, given the importance that people place on having children, analysts have used a definition that more likely would enhance the welfare of a higher number of individuals. This objection is especially significant because it focuses on the importance that individuals concede to having children. When people have a planned course of action that include having a child, waiting three years to seek treatment may not seem a desirable strategy. This is especially the case for women that have delayed pregnancy in order to pursue their careers. Thus, although 40% of couples diagnosed as infertile will conceive within three years without treatment, many may think that such a period of time is too long. Some argue that because infertility causes great mental and emotional suffering there is an obligation to relieve such suffering when the means to do so exist.[104] Therefore, in assuming the standard medical definition of 'infertility' assessors likely have increased the welfare of many individuals.

Although concerns for infertile people's suffering are laudable, this objection (that analysts have used a definition that more likely would enhance the welfare of a higher number of individuals) is incorrect because it ignores the fact that direct and indirect costs of IVF are, as we have seen, high. These expenditures, however, may be unnecessary because of the overdiagnosis of infertility. In spending money on needless treatment, states lose the opportunity to use that money for infertility prevention, prenatal care, or mother's and children's health care.[105] Nevertheless, investment in these latter fields would solve the problems of many more infertile women because the IVF success rates are very low and the treatment is expensive. Thus, funding for basic societal and medical problems would likely have a broader impact on women's ability to bear healthy children. For example,

---

104. Subcommittee on Health and the Environment, *In Vitro Fertilization-Oversight* (Washington, D.C.: U.S. Government Printing Office, 1978), p. 109.

105. See, Henifin, Equity; and World Health Organization (WHO), "Recommendations on the Management of Services for *in Vitro* Fertilization from the WHO 1990," *British Medical Journal*, 305 (July 25, 1992): 251.

improved education, health care, working conditions, and nutrition could dramatically enhance the ability of poor women and women of color to have children.[106]

1.5.4. USE OF AVAILABLE DATA

The fourth objection (that analysts require some data on the prevalence of infertility in order to arrive at their conclusions and therefore may criticism is incomplete because it does not take this requirement into account) is certainly compelling because it focuses on a practical point. Obviously, assessors need statistical information on the number of infertile people in the community in order to evaluate IVF and related procedures. Even admitting, as the British and the United States report do, that available data are far from accurate, they need some information with which to work, and they use the best accessible material. To require, for example, analysts to conduct their own investigations on infertility prevalence so that they obtain more reliable information on which to base a course of action would be excessively lengthy and expensive.

This objection, however, is questionable because it overlooks the fact that although new and more accurate research on the prevalence of infertility would likely be a desirable course of action when assessing IVF, it is not the only possible option. Certainly, doing new research on the prevalence of infertility could delay the assessors' IVF analysis. Also, funding for such research could be difficult to obtain. However, evaluators could have emphasized the inaccuracy and lack of data on infertility prevalence. Likewise, they could have accented the importance of such evidence to their analyses. For example, they could have stressed the fact that, because the definition of 'infertility' was broad, many women could be encouraged to pursue unnecessary treatment. In so doing, policy makers could have been more likely to take into account the significance of such information and the possible consequences of distinct policy options. For instance, they could have devoted resources to research the incidence of reproductive problems. Thus, decisionmakers and the public would have access to better information in which to base their decisions about infertility treatments.

---

106. See, for example, Nsiah-Jefferson, Reproductive Laws.

1.5.5. STANDARD MEDICAL DEFINITION

The fifth objection (that assessors have adopted the one year definition of 'infertility' because it is the standard medical definition, and most people seeking treatment also share it), is questionable because it begs the question. Begging the question occurs when one assumes what one is trying to prove. Thus, what proponents of this objection need to show is that the standard medical definition of 'infertility' actually represents the infertile population. If one assumes that couples would need medical intervention if they have not conceived a child after one year of regular intercourse, then one begs the question. As we have seen, there is no specific point in time at which a couple trying to conceive ceases to be fertile and become infertile. Moreover, because of the ambiguity of the standard definition (we are not sure how frequently the intercourse should occur), assessors should explain why it would be reasonable to accept the standard definition, rather than to conclude that it is adequate. They should have taken into account framing problems. Obviously, had assessors chosen a longer time frame for defining infertility, the number of individuals included in the definition would have decreased. Therefore, the framing of the infertility questions may have controlled the answers. The large number of people said to be suffering from reproductive problems dictates the seriousness of the infertility problem, which, in turn, may unnecessarily and erroneously justify the use and funding of these risky technologies.

1.6. Summary and Conclusion

The consequences for the community of undervaluing policy options such as prevention of infertility and the impacts of unnecessary use of medical technologies are important subjects meriting careful consideration. The Victorian, the British, the Spanish, and the United States reports on IVF have paid little attention to problems for the common good that could result from unnecessary use of IVF and from undervaluing preventive measures to infertility. Because these assessments influence public policies on infertility treatment, their explicit or implicit recommendations may have significant impacts on proposals that would affect not only people with reproductive problems, but also the whole community. This chapter has shown how an individualistic approach to the evaluation of infertility treatments may be, in part, responsible for an accent on IVF and related techniques to solve infertility and for an underemphasis of more conservative solutions to reproductive problems. I also have argued that evaluators of IVF have erred in their analyses because, in assuming too broad a definition of 'infertility', assessors may have encouraged people to use IVF when they do not need it. Neglecting social, ethical, and political solutions to the problem of infertility, and encouraging unnecessary use of IVF therapy may have negative consequences for the common good. Assessors might promote public policies that may put women at risk, that misallocate scarce resources, and that disregard social responsibility for the community's well being.

Developing and using new technologies without prior evaluation may have disastrous consequences for our society. Implementing a new technology after an inadequate assessment also may have adverse ethical, social, economic, and political impacts. To prevent unwanted effects, assessors should pay special attention to the implicit assumptions on which they base their evaluations. As we have seen, not to do so could jeopardize both their assessments and public welfare.

CHAPTER 6

FREE INFORMED CONSENT AND IN VITRO FERTILIZATION

1.1. Introduction

In 1978, when Lesley Brown gave birth to the first baby born though IVF, she did not know she would be the first such mother. She thought there were many others. Doctors Robert Edwards and Patrick Steptoe did not inform her that she was participating in an experiment.[1] In 1995, Ricardo Asch, an internationally acclaimed fertility specialist, and some of his colleagues at the University of California, Irvine, were accused of performing research without the consent of patients, prescribing unapproved drugs, and taking eggs and embryos from some patients without the donors' knowledge.[2]

1.2. Overview

In spite of the moral and legal requirement of obtaining informed consent from patients, misconduct such as that of Edwards, Steptoe, and Asch still occurs. In this chapter I argue that assessors of IVF have failed because their evaluations may encourage public policies that jeopardize women's rights to free informed consent. In section three I offer a brief account of the doctrine of free informed consent. Next, I present an analysis of how IVF assessors have dealt with issues of informed consent to IVF and related procedures. In section five I argue that because

1. See, for example, J. A. Scutt, "Introduction," in J. A. Scutt, (ed.), *The Baby Machine* (London: Green Print, 1990), p. 9.

2. See, for example, G. Cowley, "Ethics and Embryos," *Newsweek*, 125: 24 (June, 1995): 66-68; R. Dalton, "Fertility Pioneers Face 'Misconduct' Charges," *Nature*, 376: 6540 (August 1995): 456; and J. A. Robertson, "Eggs, Embryos, and Professional Ethics," *The Chronicle of Higher Education*, 42: 17 (January 1996): A64.

evaluators have neglected problems with disclosure of information, their analysis may compromise women's opportunities to give free informed consent. Assessors have overlooked questions of disclosure of information because (i) they have underestimated the lack of scientific evidence on IVF safety, (ii) they have undervalued difficulties with the presentation of IVF success rates, and (iii) they have overemphasized the benefits of the procedure and have downplayed the hazards. Hence, analysts have skewed the balance of pros and cons and have made it difficult for women to give informed consent to a risky treatment whose effects are uncertain. In section six I claim that assessors also have underestimated problems with voluntariness. Because evaluators have failed to analyze the social and economic conditions in which women make decisions about IVF, they have overlooked circumstances that could threaten free informed consent. Finally, I respond to some potential objections against my arguments. For example, critics may say that assessors clearly advise doctors to obtain written informed consent forms from IVF patients. They also may argue that the task of IVF assessors is to analyze IVF and related procedures, not to offer value judgments about the context in which people implement and use these technologies.

I have focused my analysis of IVF assessments on women's decisions because they are the ones who bear children, and they are the ones who have to undergo risky treatments. However, my evaluation of the importance of disclosure and the need for analyzing social factors that may hinder people's possibilities for free informed decisionmaking applies also to men's choices.

1.3. Free Informed Consent

Shortly after the Nuremberg trials, which presented horrifying accounts of Nazi experimentations on unwilling human subjects, the issue of informed consent began to receive attention.[3] The first sentence of the Nuremberg Codes states that the voluntary consent of human subjects in

3. See, for example, Ruth Faden and Tom Beauchamp, *A History and Theory of Informed Consent* (New York: Oxford University Press, 1986) (hereafter cited as Faden and Beauchamp, *History*).

research is absolutely essential. At Helsinki in 1964, the World Medical Association made consent of patients and subjects a central requirement of ethical research.[4] Since then virtually all prominent medical and research codes as well as institutional rules of ethics dictate that physicians and investigators obtain the free informed consent of patients and subjects prior to any substantial intervention.[5] Procedures for free informed consent have several functions such as the protection of patients and subjects from harm or the promotion of medical responsibility in interactions with patients and subject. Their more fundamental goal is, however, to enable autonomous choices.[6]

The received approach to the definition of informed consent has been to specify the elements of the concept. Legal, regulatory, medical, psychological, and philosophical literature tend to analyze informed consent in terms of the following elements[7]: (1) disclosure, (2) understanding, (3) voluntariness, and (4) competence. Thus, one gives free informed consent to an intervention if and only if one is competent to act, receives a thorough disclosure about the procedure, understands the disclosure, acts voluntarily, and consents to the intervention. Disclosure refers to the necessity of professionals' passing on information to

---

4. See, for example, Jay Katz, *Experimenting with Human Beings* (New York: Russell Sage Foundation, 1972), pp. 312-13; Faden and Beauchamp, *History*, ch. 5; and W. T. Reich, ed., *Encyclopedia of Bioethics* (New York: Simon & Schuster Macmillan, 1995), Appendix (hereafter cited as Reich, *Encyclopedia*).

5. See, for example, Faden and Beauchamp, *History*, ch. 3; and Reich, *Encyclopedia*.

6. See, for example, Faden and Beauchamp, *History*; and T. L. Beauchamp and J. F. Childress, *Principles of Biomedical Ethics*, 4th ed. (New York: Oxford University Press, 1994), ch. 3; hereafter cited as Beauchamp and Childress, *Principles*.

7. See, for example, U.S. National Commission for the Protection of Human Subjects of Biomedical and Behavioral Research, *The Belmont Report: Ethical Principles and Guidelines for the Protection of Human Subjects of Research* (Washington D.C.: U.S. Government Printing Office, 1978); A., M., and L. H. Roth, "What We Do and Do Not Know About Informed Consent," *Journal of the American Medical Association*, 246:21 (1981): 2473-2477; U.S. President's Commission for the Study of the Ethical Problems in Medicine and Biomedical and Behavioral Research, *Making Health Care Decisions: A Report on the Ethical and Legal Implications of Informed Consent in the Patient-Practitioner Relationship* (Washington, D.C.: U.S. Government Printing Office, 1982); Faden and Beauchamp, *History*; and Beauchamp and Childress, *Principles.*, ch. 3. The analysis of informed consent relies mainly on Faden and Beauchamp, *History*, part III.

decision-makers and possible risk victims. The professionals' perspectives, opinions, and recommendations are usually essential for a sound decision. They are obligated to disclose a core set of information, including (1) those facts or descriptions that patients usually consider material in deciding whether to consent of refuse to the intervention, (2) information the doctors believe to be material, (3) the professionals' recommendations, (4) the purpose of seeking consent, and (5) the nature and limits of consent as an act of authorization.

Understanding may be the most important component for free informed consent. It requires professionals to help potential risk victims overcome illness, irrationality, immaturity, distorted information, or other factors that can limit their grasp of the situation to which they have the right to give or withhold consent. Thus people understand if they have acquired pertinent information and justified, relevant beliefs about the nature and impacts of their actions. This understanding need not be complete, because a substantial grasp of central facts is generally sufficient. Normally, diagnoses, prognoses, the nature and purpose of the intervention, alternatives, risks, benefits, and recommendations are essential. Patients or subjects also need to share an understanding with professionals about the terms of the authorization before proceeding. Unless agreement exists (about the crucial features of what the patients authorize) there can be no assurance that they have made autonomous decisions. Thus, even if doctor and patients use a word such as 'ovulation induction', their interpretations could be totally different if standard medical definitions have no meaning for the patients.

Another element of informed consent is voluntariness, or being free to act in giving consent. It requires that the subjects act in a way that is free from manipulation and coercion by other persons. Coercion occurs if and only if one person intentionally uses a credible and serious threat of harm or force to control another.[8] Manipulation is convincing people to do what the one wants by means other than direct coercion or rational persuasion. One important form of manipulation in health care is informational manipulation, a deliberate act of handling information that

---

8. See, for example, R. Nozick, "Coercion," in S. Morgenbesser, P. Suppes, and M. White (eds.), *Philosophy, Science, and Method: Essays in Honor of Ernest Nagel* (New York: St. Martin's Press, 1969), pp. 440-72.

alters patients' understanding of the situation and motivates them to do what the agent of influence plans. Withholding evidence, misleading exaggerations of benefits are all instances of manipulation inconsistent with voluntary decisions. The way in which doctors present information by tone of voice, by framing information positively (the therapy is successful most of the time) rather than negatively (the therapy fails in forty percent of the cases) can manipulate patients' perceptions and, therefore, affect understanding.

The criterion of competence refers to the patients' abilities to perform a task. Thus patients or subjects are competent if they have the ability to understand the material information, to make a judgment about the evidence in light of their values, to intend a certain outcome, and freely to communicate their wishes to the professionals.

1.4. Free Informed Consent and In Vitro Fertilization Assessments

Certainly, the four committees defended the importance of obtaining informed consent from patients. For example, the Victorian Committee recommended that "each couple seeking admission to and IVF programme must be provided with clear and comprehensive information about the programme. Each couple must be advised of any risks involved in the programme and of its likely duration and given the most accurate information about the prospects of success. This will then permit each couple to express in writing their free and informed consent to participate in the programme."[9]

The members of the British committee also "feel it to be very important that time and consideration be devoted to explaining fully to prospective patients and, where necessary, to their partners the details of any infertility treatment that they undergo."[10] British IVF assessors

---

9. See, Committee to Consider the Social, Ethical, and Legal Issues Arising from In Vitro Fertilization, *Interim Report* (Victoria: Victorian Government Printer, 1982), p. 23 (hereafter cited as Victorian Report).

10. M. Warnock, *A Question of Life. The Warnock Report on Human Fertilization and Embryology* (Oxford, UK: Oxford University Press, 1985), p. 16(hereafter cited as Warnock Report).

consider that "no such treatment should be undertaken without the fully informed consent of the patient and this should, in the case of more specialized treatment, normally be obtained in the presence of someone not associated with the performance of the procedures."[11] Thus, the members of the committee "recommend that in the case of more specialized forms of infertility the consent in writing of both partners should be obtained, wherever possible, before treatment is begun, as a matter of good practice. Any written consent should be obtained on an appropriate consent form."[12] British evaluators also think that it is "desirable that the process of explaining and describing prospective treatment should be embarked upon as far in advance as possible so that both partners have plenty of time to discuss and consider treatment."[13]

As for the Spanish Commission, it recommends that professionals give "sufficient information and advice on possible consequences of these techniques, such as risks and benefits, to those who wish to use assisted conception techniques, or wish to be donors of gametes and human embryos."[14] Commissioners propose that "information and advice be extended to legal, biological, ethical, and economic issues."[15] The acceptance of IVF and related procedures must appear, according to the Spanish assessors, "in a consent form, filled and signed by the patients and/or donors."[16] Also, evaluators require that "women must freely and responsibly sign the consent form in order to undergo infertility treatment with IVF and related procedures and they may ask to terminate the process at any time."[17] In cases of married or stable couples, and when the treatment requires donation of sperm or eggs, both women and men must give written consent to the procedure.[18]

---

11. Warnock Report, p. 16.
12. Warnock Report, pp. 16, 82.
13. Warnock Report, p. 16.
14. Comisión Especial the Estudio de la Fecundación *In Vitro* y la Inseminación Artificial Humanas [Special Commission for the Study of Human *In Vitro* Fertilization and Artificial Insemination], *Informe* [Report] (Madrid: Gabinete de Publicaciones, 1987), p. 114 (hereafter cited as Spanish Commission).
15. Spanish Commission, p. 114.
16. Spanish Commission, p. 114.
17. Spanish Commission, p. 114.
18. Spanish Commission, p. 114.

Finally the United States report mentions the importance of "the rights of patients who are being treated for infertility to be appropriately informed about the research aspects of their treatment."[19] It also argues that because individuals with infertility problems are an extremely vulnerable population, their ability to give free informed consent is to some extent always compromised. Thus, U.S. assessors affirm, "it is particularly important that care be taken to carefully inform infertile couples when new reproductive technologies are suggested as possible methods of treatment."[20] Evaluators also consider essential "for the patient to know the truth about a specific treatment and likelihood of success of a given effort."[21]

The goal of protecting informed consent, however, may be compromised because, in all four reports, assessors have underestimated problems with some of the components of informed consent, namely disclosure and voluntariness. In the next section I show that IVF assessors have failed to address adequately problems with disclosure of information. In section five, I argue that IVF evaluators have undervalued problems of voluntariness.

1.5. Failing to Promote Free Informed Consent: Disregarding Problems with Disclosure

The most infamous case of prolonged and knowing violation of free informed consent was a U.S. Public Health Service study initiated in the early 1930s and ended in 1973. The stated purpose of the Tuskeegee Study was to compare the health and longevity of an untreated syphilitic population with a nonsyphilitic but otherwise similar population.[22]

---

19. Office of Technology Assessment, *Infertility: Medical and Ethical Choices* (Washington, D.C.: U.S. Government Printing Office, 1988), p. 211 (hereafter cited as OTA, *Infertility*).

20. OTA, *Infertility*, p.211.

21. OTA, *Infertility*, p. 211.

22. See, Committee on Labor and Human Resources, *Biomedical Ethics and U.S. Public Policy* (Washington, D.C.: U.S. Government Printing Office, 1994), p.3; J. H. Jones, *Bad Blood* (New York: Free Press, 1993), revised edition (hereafter cited as Jones, *Bad Blood*); and Faden and Beauchamp, *History*, pp.165-67.

The subjects in the Tuskeegee case knew neither the name nor the nature of their disease. No one told them that they were participants in a nontherapeutic experimentation. They also were misinformed that painful procedures would be part of the treatment. Moreover, researchers systematically prevented untreated patients from receiving available treatment.[23]

In 1972, after reporter Jean Heller published an account of the study, the Department of Health, Education, and Welfare (DHEW) appointed an ad hoc advisory panel to review the Tuskeegee research. The panel found that neither the DHEW nor any other government agency had a uniform or adequate policy for reviewing experimental procedures of securing subjects' consent. The panel recommended that the Tuskeegee study be terminated at once, and that the remaining subjects be given the necessary care. With respect to obtaining adequate informed consent, the panel recommended that a national committee be appointed with the mission of educating subjects about their rights and ensuring the quality of consent statements.[24]

Because one of the main goals of the IVF assessments under analysis is to advise governments in the selection of sound public policies (regarding infertility treatments), evaluators should have paid special attention to factors that could compromise women's ability to give free informed consent. Following is a discussion of how assessors have failed to address adequately problems with disclosure of information. First, evaluators have underestimated the lack of empirical data on IVF risks and effectiveness. Second, they have overlooked difficulties with the presentation of IVF success rates. Third, they have downplayed IVF hazards and have overemphasized its benefits.

---

23. See Jones, *Bad Blood*; and Faden and Beauchamp, *History*, pp.165-67.

24. See *Final Report of the Tuskeegee Syphilis Study Ad Hoc Panel* (Public Health Service, April 28, 1973); Jones, *Bad Blood*; and Faden and Beauchamp, *History*, pp.165-67.

1.5.1. UNDERESTIMATING LACK OF DATA ON IN VITRO FERTILIZATION RISKS AND EFFECTIVENESS

As we have seen in chapter four, the Victorian, British, Spanish, and the United States committees undervalue the insufficiency of evidence on the effectiveness and safety of IVF and related procedures.[25] They have underestimated the scarcity of evidence concerning the long-term effects (such as ovarian and breast cancer) of ovulation drugs on women. Similarly, they have failed to take into account adequately that little or no follow-up information exists on what happens to children and women after pregnancy with IVF or related procedures. Furthermore, evaluators also have neglected the fact that studies on the psychological and emotional impacts of these techniques on women and their children are scarce. It is true that, as I have said, assessors recognize the need for improvement of these procedures.[26] Nevertheless, neither the Victorian, the British, nor the Spanish reports pay enough attention to the lack of research on physical and psychological risks associated with IVF. Only the United States assessment mentions the insufficiency of investigations on maternal health consequences and anomalies in offspring.[27] It does not concede, however, particular relevance to this lack of information.

In underestimating the insufficiency of good investigations (i.e., studies with control groups, large sample sizes, long-term follow up), IVF assessors have failed to provide adequate information about these problems to policymakers and to those considering using these technologies. Because they have inadequately addressed problems with disclosure, evaluators may have compromised women's opportunities to consent to treatments whose effects are uncertain and potentially dangerous. Obviously, requiring free and informed consent would be ineffective without information's being available on hazards and benefits of IVF and related techniques and without disclosure about the uncertain impacts of the treatment. Lack of, or faulty, information may seriously hinder people's abilities to make informed choices. If women undergoing IVF do not know, for instance, their real chances of having a child, the

---

25. See Chapter Four for references.
26. See Chapter Four.
27. OTA, *Infertility*, pp. 302-303.

risks associated with ovulation hormones, or the hazards related to multiple-birth pregnancies, they cannot be said to be informed. Hence, if women do not have information that is relevant for their decisions, then they cannot give free informed consent.

Had the IVF commissioners specifically acknowledged in their assessments the lack of scientific information on IVF and other infertility therapies, they could have proposed recommendations that would more likely preserve the free and informed consent of women undergoing treatment. For instance, they could have required that researchers gather more data on short and long-term effects of drugs and procedures. Certainly, a more comprehensive evaluation of such outcomes may result in more informed patients, who would be capable of making appropriate decisions related to infertility treatments.[28] Also, assessors could have demanded that more investigation be completed on the effectiveness of IVF for some reproductive problems--i.e., unexplained infertility or male-related factors--before doctors offered IVF as a therapy for these conditions. Until researchers collected such information, commissioners could have advised, for example, that doctors offer IVF only in the context of experimentation on clinical trials. Because scientific evidence has not proved IVF to be effective for conditions others than tubal blockage, recommendations limiting the use of this procedure could likely prevent unethical experimentation on women undergoing IVF treatment for those other indications (male factors or unexplained infertility). Furthermore, had the committees adequately recognized the importance of the lack of evidence on different aspects of IVF, they particularly could have required that doctors inform women about what is known and what is not known about hazards related to IVF. Certainly, gaps in research, evaluation, and record keeping may make it impossible to provide full information in all areas. However, where professionals cannot meet information needs, commissioners could demand that materials written for patients should state explicitly that reliable data are

---

28. J. Jarrel, J. Seidel, and P. Bigelow, "Adverse Health Effects of Drugs Used for Ovulation Induction," in *New Reproductive Technologies and the Health Care System. The Case for Evidence-Based Medicine*, Royal Commission on New Reproductive Technologies (Ottawa, Canada: Canada Communications Group, 1993), p. 523.

not available. Thus, women would be able to give free informed consent to risky treatments whose effects are uncertain.

1.5.2. UNDERVALUING DIFFICULTIES WITH IN VITRO FERTILIZATION SUCCESS RATES

A second problem that exhibits the commissioners' failure to foster free informed consent is related to the data on IVF success rates. Obviously, adequate knowledge about the real chances of having a child through IVF is relevant when making decisions to undergo treatment. Nevertheless, IVF assessors have failed to address satisfactorily problems with the presentation of IVF success rates that could prevent women from giving free informed consent.

Although IVF clinics usually publish up to 20% success rates, some of them have not produced even one live child.[29] As we have seen in chapter three, IVF consists of a series of treatment phases. A woman may drop out of any stage because the treatment failed. Therefore, when estimating success rates, using embryo-transfer cycles as the denominator will increase the success rate, while using all started treatments will decrease it. Similarly, when doctors use clinical pregnancies as the measure of success, then they count as success of the treatment chemical pregnancies (changes in the woman's hormonal levels), ectopic pregnancies, miscarriages, and preterm births.[30] In this case, professionals

---

29. See, for example, Subcommittee on Regulation and Business Opportunities, *Consumer Protection Issues Involving in Vitro Fertilization Clinics* (Washington, D.C.: U.S. Government Printing Office, 1988) (hereafter cited as SRB, *Consumer*, 1988); Subcommittee on Regulation, Business Opportunities, and Energy, *Consumer Protection Issues Involving in Vitro Fertilization Clinics* (Washington, D.C.: U.S. Government Printing Office, 1989) (hereafter cited as SRB, *Consumer*, 1989); Subcommittee on Health and the Environment, *Fertility Clinic Services* (Washington, D.C.: U.S. Government Printing Office, 1992) (hereafter cited as SHE, *Fertility*); and Rowland, *Living Laboratories* (Bloomington: Indiana University Press, 1992), p. 44-49 (hereafter cited as Rowland, *Laboratories*).

30. See, for example, J.B. Russell and A.H. DeCherney, "Helping Patients Choose and IVF Program," *Contemporary Obstetrics and Gynecology*, 30 (1987): 145-49; SHE, *Fertility*; SRB, *Consumer*, 1988; and SRB, *Consumer*, 1989; Rowland, *Laboratories*,

inflate the successful results of the treatment. Because professionals define "success" in loose ways, IVF success rates appearing in the media are misleading.[31] Such statistics do not specify the numerator and denominator of the rate. They fail to reveal, for example, if the denominator is the number of ovarian stimulations that the doctors have performed, the quantity of eggs retrieved, the number of eggs fertilized, or the total of implanted embryos. Also, in these statistics there is no information about the number of menstrual cycles that were necessary to achieve a pregnancy. Likewise, the statistics published in the media fail to disclose client characteristics, although success rates vary depending on the underlying cause of infertility. For example, women treated because their partners have reproductive problems are much less likely to benefit from IVF compared with women having tubal blockage. Success rates also diminish for older women and for women who have remained infertile for longer periods of time.[32]

Because patients, clinics, and practitioners define "success" in IVF treatment differently, prospective patients cannot assess their likelihood of having a child through the use of IVF. Without clearly defined and universally employed definitions of 'success', it is impossible to know what is being compared, i.e., whether it is successful fertilization, implantation, a clinical pregnancy, or a live birth. Without this information, meaningful evaluation and decisionmaking is impossible. Success rates quoted without a frame of reference mislead people about their probability of having a child.

Despite the importance of accurate information on IVF success rates, as a basis for giving free informed consent, the Victorian Report does not mention this problem. The British Committee recognizes that it is difficult "to give an estimate of the success of the technique [IVF] because of differing methods of measuring success and also because rates

---

pp.44-49; L. S. Wilcox *et al.*, "Defining and Interpreting Pregnancy Success Rates for In Vitro Fertilization," *Fertility and Sterility*, 60: 1 (1993): 18-25.

31. See, for example, M. R. Soules, "The In Vitro Fertilization Pregnancy Rate: Let's Be Honest with One Another," *Fertility and Sterility,* 43: 4 (1985): 511-513.

32. See, for example, F. J. Stanley and S. M. Webb, "The Effectiveness of In Vitro Fertilization: An Epidemiological Perspective, in *Tough Choices*, eds., P. Stephenson and M. G. Wagner (Philadelphia: Temple University Press, 1993), pp.62-72.

vary between centers."[33] The Committee, however, does not make any specific recommendations on how to deal with the problem of estimating accurate success rates for IVF treatments. The Spanish report affirms that "in a well qualified IVF program 20% of women undergoing IVF achieve a pregnancy."[34] There is no acknowledgment of any problem related to the estimates of the success rates of the procedure. The Committee makes no suggestions on how best to present such rates. The United States study acknowledges that "interpreting effectiveness data in the field of infertility treatment is difficult and controversial.[35] Reasons for this difficulty are, according to U.S. assessors, a problem with determining what particular actions may have been responsible for a pregnancy and the considerable variation in which clinics report pregnancy rates.[36] However, evaluators do not relate this lack of accurate success estimates to epistemological and ethical problems with free informed consent. However, as I have said, if women have misleading information about the success of IVF they cannot be said to be informed. Therefore, they cannot give free informed consent.

Had assessors recognized the importance of accurate information on the benefits of IVF, they could have suggested that clinics offer these data in a uniform way or that they specify what they take as measures of success. Because knowing the chances of having a child is essential information for deciding whether to undergo IVF treatment, advising policymakers of the need for accurate presentation of IVF success rates likely would have helped to protect women's rights to free informed consent.

But informed consent issues are not the only important factors when evaluating IVF success rates. Records on outcomes are also necessary to guide practice and to help develop research and public policy. They are needed to allow analyses that give rise to better decisions about whether doctors should offer IVF as a therapy, whether they should abandon its use for certain conditions, whether they should consider the treatment as experimental, or whether governments should spend public money on it.

---

33. See Warnock Report, p.33.
34. See Spanish Commission, p. 107.
35. See OTA, *Infertility*, p. 146.
36. See OTA, *Infertility*, p. 146.

1.5.3. SKEWING THE RISK-BENEFIT BALANCE

Assessors have failed to address adequately problems with disclosure not only when underestimating the lack of evidence on IVF safety and effectiveness or when undervaluing difficulties in the presentation of IVF success rates. They also have failed to tackle disclosure issues satisfactorily when underestimating the risks of this procedure and emphasizing its benefits. In this section I argue that IVF evaluators have skewed the balance of pros and cons and, therefore, they might have encouraged public policies that make it difficult for women to give free informed consent. I suggest that social norms about motherhood may have played a role in their downplaying IVF risks and overemphasizing its benefits.

As I have shown in chapter four, assessors have failed to disclose adequately relevant data on known risks associated with IVF and related procedures. They have underestimated the information on short- and long-term hazards associated with ovulation-induction hormones. Especially important is the lack of information in their evaluations on the existing evidence relating reproductive cancers and hormones. As I have mentioned in chapter four, a substantial body of experimental, clinical, and epidemiological evidence indicates that hormones play a major role in the development of several human cancers.[37] Data indicate that hormone-related cancers account for more than 30% of all newly diagnosed female cancer in the United States. Increases in cancer rates are related to greater use of hormones for contraception, and for menstruation and menopause problems.[38] In spite of the significance of this information, IVF evaluators fail to mention it.

---

37. See, for example, S. Fishel and P. Jackson, "Follicle Stimulation for High-Tech Pregnancies: Are We Playing It Safe?" *British Medical Journal,* 299 (1989): 309-11; and P. Stephenson, "Ovulation Induction During Treatment of Infertility: An Assessment of the Risks," in *Tough Choices,* eds., P. Stephenson and M. G. Wagner (Philadelphia: Temple University Press, 1993), pp.105-107.

38. See, for example, H.P. Schneider and M. Birkhauser, "Does Hormone Replacement Therapy, Modify Risks of Gynecological Cancers?" *Int. J. Fertil. Menopausal Stud.,* 40, suppl. 1 (1995): 40-53; T. J. Key, "Hormones and Cancer in Humans," *Mutat Res,* 333: 1-2 (1995): 59-67; F. Berrino *et al.,* "Serum Sex Hormone Levels after Menopause and Subsequent Breast Cancer," *J Natl Cancer Inst,.* 88: 5, (1996): 291-6.

In chapter four I also showed that assessors have underestimated the hazards (postoperative infections, punctures of an internal organ, hemorrhages, ovarian trauma, and intrapelvic adhesions) associated with the procedures that doctors normally use to obtain women's eggs, i.e., laparoscopy and ultrasound-guide oocyte retrieval. Similarly, they have undervalued the need for information on the risks (perforation of organs, ectopic pregnancies, and multiple gestation) related to the implantation of embryos or gametes into women's bodies. In failing to disclose adequately relevant information on IVF hazards, assessors have made IVF appear as less risky than it may be. Thus, they have downplayed the significance of IVF risks.

At the same time that evaluators have minimized the importance of IVF dangers, when presenting this procedure as a successful therapy, assessors have also overemphasized its benefits. For instance, the Victorian report stresses the success of the procedure by narrating different "breakthroughs" leading to improved success rates.[39] Victorian assessors mention major success achievements "when culture media which enabled 90% of the ripe eggs to be fertilized in the laboratory were produced; [when] further improved techniques for the collection of ripe eggs were developed."[40] The final advance was, evaluators say, "the discovery that success rates were improved if more than one embryo was able to be transferred into the uterus."[41] They fail to cite possible risks associated with these "breakthroughs." Similarly, the British committee offers as an illustration of the success of IVF the results of treatments at Bourn Hall Clinic in the United Kingdom. Such outcomes are an indication, assessors declare, that "the technique had now passed the research stage and can be regarded as an established form of treatment for infertility."[42] They do not mention results in other clinics that could be less successful. The Spanish report also accentuated the effectiveness of IVF by mentioning that the success rates of the procedure were very similar to those of natural conception.[43] Assessors affirm that in a good

---

39. See Victorian Report, pp. 8-10.
40. See Victorian Report, pp. 8-10.
41. See Victorian Report, pp. 8-10.
42. See Warnock Report, p.34.
43. See Spanish Commission, p. 107.

IVF program "20% of women undergoing IVF achieve a pregnancy; that means, 1 of every 5 women, which with only one attempt, gives a rate similar to that of natural gestation (30%)."[44] Finally, the United States assessment stressed IVF benefits by giving the rates of success at the most expert clinics.[45] They say that "the chances of achieving a successful pregnancy in the hands of the most expert practitioners are estimated to be about 15 to 20 per cent for one complete IVF cycle."[46] They fail to mention how many "expert clinics" exist.

Overemphasizing IVF benefits is problematic because it may mislead people into believing IVF to be more successful than it is. Obviously, misleading information may jeopardize women's abilities to give free informed consent.

Social norms may have played a role in downplaying IVF risks and in emphasizing its benefits.[47] For example, the four reports stressed the importance of having a child and forming a family.[48] All of them have accentuated the pain of being involuntarily childless, and all of them consider IVF an acceptable treatment for infertility. Thus, because of cultural norms about the importance of motherhood, and scientific uncertainty about IVF risks and effectiveness, assessors might have tolerated a relatively high level of risk when interpreting data on IVF. As a result they might have neglected the hazards associated with IVF treatment when analyzing this procedure. Such risk-tolerant values, used in interpreting data and disclosing risks, would poorly protect women's welfare and ability to make informed choices.

The presupposition that cultural norms may have played a role in downplaying IVF risks appears plausible when analyzing other cases in which social practices have influenced risk evaluation. An example of the effects of normative commitments on the interpretation of scientific results appears in the case of contraception. According to some studies,

---

44. See Spanish Commission, p. 107.

45. OTA, *Infertility*, p. 293,305.

46. OTA, *Infertility*, p. 305.

47. See, for example, L. A. Parker, "Beauty and Breast Implantation: How Candidate Selection Affects Autonomy and Informed Consent," *Hypatia*,. 10: 1 (Winter 1995): 183-201.

48. See, for example, Victorian Report, p. 4; Warnock Report, p.8; Spanish Commission, p.53; OTA, *Infertility*, p. 37-8.

doctors' interpretations of empirical data concerning oral contraception revealed that those who viewed it as immoral believed oral contraception to be much more unsafe and ineffective than those who viewed it as moral.[49] Similarly, given the importance the assessors place on having children and forming a family, cultural views might have influenced their evaluation of IVF risks and benefits.

Whatever the reasons for emphasizing IVF benefits and underestimating its risks, the question is that (in stressing the success of the procedure and downplaying its hazards), these IVF committees have skewed the balancing of pros and cons. Hence, assessors have failed to consider adequately the fact that accurate disclosure of benefits and risks is essential when giving free informed consent. Hence, they may have encouraged public policies that interfere with women's rights to free informed consent.

1.6.Failing to Promote Free Informed Consent: Disregarding Problems with Voluntariness

Discussions about appropriate standards for professional disclosure have generated more controversy than any other issue about informed consent.[50] However, remaining independent of controlling influences is equally important for free and informed decisionmaking. Certainly many of people's life decisions take place in contexts fraught with social demands, competing claims and interests, and expectations. Some of these influences on individuals are unavoidable and may be desirable (influences from a loved one), but others may interfere or impede voluntary choices. In this section I argue that assessors of IVF have erred in their evaluations because, in failing to analyze adequately the social context in which IVF exists, they may have encouraged public policies that hinder women's opportunities to give free informed consent.

As I said in section 2.1, the voluntariness to act is important for free informed consent. Thus people's actions are voluntary when they are free

---

49. See, for example, R. M. Veatch, *Value-Freedom in Science and Technology* (Missoula, MT: Scholars Press, 1976).

50. Faden and Beauchamp, *History*, chs. 2 and 4.

from controlling influences. However, when a particular influence becomes controlling (in a sense that prevents voluntary actions) may be difficult to know. Individuals are subjectively influenced in different ways, some people being more resistant to specific influences than others. For example, some persons find it difficult to resist the standard idea of beauty dominant in our society, and they may try impossible diets in order to achieve such an ideal. Other people, however, may judge aesthetic norms easily resistible. Situations of dependency and vulnerability also may alter the way in which controlling influences affect people. Thus individuals in situations of illness or of extreme economic need may be more easily controllable than otherwise. The goal of this section is to criticize IVF assessments for ignoring important social, political, and economic situations that may prevent women form acting voluntarily. Assessors' disregard of such situations is problematic because their analyses may promote public policies that restrict women's opportunities to exercise their rights to free informed consent.

Some examples may serve to illustrate cases in which a decision may be far from voluntary because of very severe social, political, or economic restrictions. In countries like China and India, where the number of children is tightly limited, women are under extremely powerful social and economic constraints to have sons.[51] Sons can help out in the fields, can take care of their parents in their old age, and will continue the family line. On the contrary, the birth of daughters appears to be a misfortune because they have scarce earning power, they will leave the family once they married, and they may need large dowries. Women who do not produce sons may be abused, neglected, or abandoned. Under these circumstances, a woman may find it impossible to refuse to abort a female fetus, even if the prospect insults her moral convictions. Thus, if social and economic conditions have actively

51. See, for example, M. A. Warren, *Gendercide: The Implications of Sex-Selection* (Totowa, NJ: Rowman and Allanheld, 1985); A. Kroeber, "Preventing Women from Being Born," *The Progressive*, 52: 12 (Dec 1988) : 14-16; A. R. Holder and M. S. Henifin, "Selective Termination of Pregnancy," *Hastings Center Report*, 18: 1 (Feb-March 1988): 21-23; E. Bumiller, *May You Be the Mother of A Hundred Sons: A Journey among the Women of India* (New York: Fawcett Columbine, 1990); J. Butler, *Bodies that Matter: On the Discursive Limits of Sex* (New York: Routledge, 1993); and G. Weiss, "Sex-Selective Abortion: A Relational Approach," *Hypatia*, 10: 1 (Winter 1995): 202-217.

impeded all her other options, and although it may be difficult to fix the blame on a specific individual, her consent to have an abortion would be short of being a case of free informed consent.

Fetal testing constitutes another example in which social, political or economic circumstances may limit women's opportunities to give free informed consent.[52] The current ability to diagnose diseases (i.e., Downs Syndrome, Tay Sachs, spina bifida) in utero raises many morally and politically problematic issues, if for no other reason than that prenatal testing suggests the option of aborting fetuses that are at risks of having a disease or disability. Prenatal diagnosis can confront prospective parents, and especially mothers, with difficult decisions about whether to have selective abortions. Women have to make these decisions in coercive social and economic circumstances that can render the option to abort a nonvoluntary one. For example, in a social context where inadequate social and medical services are available for raising a child with a disability, and where women are seen as responsible for raising the child, their decisions about fetal testing and abortion may be far from being voluntary. The unavailability of adequate options may endanger conditions for free informed consent.

Analyzing social, economic, and political conditions in which women make their decisions is also important when evaluating IVF and related technologies because such conditions might obstruct voluntary choices. The implementation and use of these procedures may have the potential to restrict women's voluntary decisions. Therefore, IVF assessments, as fundamental mechanisms in the guidance of public policy, should have paid attention to the context in which women are using these technologies. For instance, social pressures on women to become mothers may make it incredibly difficult for those with infertility problems to forego the use of any technology such as IVF that promises

---

52. See, for example, R. Faden, "Autonomy, Choice, and the New Reproductive Technologies: The Role of Informed Consent in Prenatal Genetic Diagnosis," in *Women and New Reproductive Technologies*, eds., J. Robin and A. Collins (Hillsdale, NJ: Lawrence Erlbaum Associates, 1991), pp. 37-47; A. Lippman, "Prenatal Genetic Testing and Screening: Constructing Needs and Reinforcing Inequities," *American Journal of Law and Medicine* , 17 (1991): 15-50; B. K. Rothman, "The Tentative Pregnancy: Then and Now," *Fetal Testing and Therapy*, 8, supp.1 (1993): 60-63.

a chance to deliver a child.[53] Women with reproductive problems (or whose partners are infertile) may feel pressured to use any available means to have a child. Thus, in the same way that avoiding unnecessary obstetric interventions, such as cesarean sections, is now extremely difficult for many women, it may become arduous for women with reproductive difficulties to escape from hazardous and needless treatments in the process of attempting conception.

Also, in a context where women are seen predominantly as mothers, an emphasis on IVF and related technologies may have the effect of pressuring them to use these procedures when they are faced with reproductive problems. Furthermore, because only women can have children, any reproductive technology may have a direct effect on their position in society. Stressing the role of motherhood could lead to a diminution of consideration for women's other roles.

Neglecting an adequate analysis of the social context in which IVF and associated procedures exist may also have blinded IVF evaluators to issues of power imbalance between men and women that may hinder women's opportunities to give free informed consent. The cases of fertile women undergoing IVF because their partners are infertile may illustrate this problem. As I mentioned in previous chapters, men are responsible for about half of the cases of reproductive problems. Nevertheless, there is still little knowledge of the causes and treatment of male infertility. Thus, if infertile men want to have a child with their partners, women are the ones who have to undergo medical treatment.[54] Hence, instead of

---

53. See, for example, R. Arditti, R. D. Klein, and S. Minden, *Test-Tube Women. What Future for Motherhood* (London: Pandora Press, 1984); A. Rich, *Of Woman Born. Motherhood as Experience and Intitution* (New York: Norton, 1986); M. Stanworth, ed., *Reproductive Technologies. Gender, Motherhood, and Medicine* (Minneapolis: University of Minnesota Press, 1987); C. E. Miall, "The Stigma of Involuntary Childlessness," *Social Problems*, 33 (1989): 268-82; B. K. Rothman, *Recreating Motherhood. Ideology and Technology in a Patriarchal Society* (New York: W.W. Norton & Company, 1990); A. Phoenix, A. Woollett, and E. Lloyd, eds., *Motherhood. Meanings, Practices, and Ideologies* (London: Sage, 1991); M. S. Ireland, *Reconceiving Women* (New York: The Guilford Press, 1993), and Jackson R., *Mothers Who Leave* (London: Pandora, 1994).

54. See, for example, J. D. McConnell, "Diagnosis and Treatment of Male Infertility," in *Textbook of Reproductive Medicine*, eds., in B. R. Carr and R. E. Blackwell (Norwalk, Connecticut: Appleton & Lange, 1993), pp. 453-468; Colin, D. M., "Clinical Male

trying to overcome male problems with better and more research, professionals have routinized doctoring women to the degree that treating them for diverse male conditions, i.e., low sperm count, is now acceptable and habitual. In such cases, a perfectly healthy, fertile woman will undergo long and hazardous treatments as a way to overcome male-related disorders.[55] Thus, independently of who has the infertility problem, women are the ones who often bear the burden of medical interventions.

Unequal participation in the medical and research professions by women and by members of racial and other minority groups also shows an imbalance of power that may prevent women from giving free informed consent to IVF and related procedure. Researchers must understand the effects of what they do. To do so, they may need the insights of those who are personally affected by the new reproductive technologies.[56] When disregarding the social context, assessors might neglect the impact that women's representation at every level of research and decisionmaking can have for informed consent issues. A scientific

---

Infertility. The Choice of Approaches for Pregnancy," *Reproduction, Fertility, and Development*, 6:1 (1994): 13-18; Cunningham, G.R., "Male Factor Infertility," in *Reproductive Medicine and Surgery*, eds., E. E. Wallach and H. A. Zacur (St, Louis, Missouri: Mosby, 1994), pp. 399-414; R. D. Kempers, "Where Are We Going?," *Fertility and Sterility*, 62:10 (1994): 686-689; J. S. Sherman, "A Modern View of Male Infertility," *Reproduction, Fertility, and Development*, 6:1 (1994): 93-104; and N. E. Skakkebaek, A. Giwercman, and D. de Kretser, "Pathogenesis and Management of Male Infertility," *The Lancet*, 343:8911 (1994): 1473-1478.

55. See, for example, R. B. Meacham, and L. I. Lipshultz, "Assisted Reproductive Technologies for Male Factor Infertility," *Current Opinion in Obstetrics and Gynecology*, 3:5 (1991): 656-661; Gordts, S. *et al*, "Role of Assisted Fertilization Techniques in the Management of Male Infertility," *Contracep Fertil Sex* 21:10 (1993): 695-700; D. Royere, "Assisted Procreation for Male Indication," *Rev. Prat.*, 43:8 (1993): 981-986; Y. S. Carmeli and D. Birenbaum-Carmeli, "The Predicament of Masculinity: Towards Understanding the Male's Experience of Infertility Treatments," *Sex Roles*,. 30: 9-10 (1994): 663-77; and M. Sigman, "Assisted Reproductive Techniques and Male Infertility," *The Urologic Clinic of North America*, 21:3 (1994): 505-515.

56. See, for example, H. E. Longino, "Knowledge, Bodies, and Values. Reproductive Technologies and Their Scientific Context," in *Technology and the Politics of Knowledge*, eds., A. Feenberg and A. Hannay (Bloomington: Indiana University Press, 1995), pp. 195-210.

community intentionally inclusive of those affected by its work may reshape procreative technologies or change the emphases of research in ways that would increase the ability of women to make free and informed decisions.

Assessors of IVF have then failed because their evaluations have neglected the fact that an appreciation--of the economic, social, and psychological pressures faced by patients, and an awareness of the ways that reproductive research may wrongly exploit those pressures--is equally essential for free informed consent issues. Inadequate evaluation of these pressures may encourage decisionmakers to propose policies that may restrict women's opportunities to give free informed consent. Had IVF evaluators recognized the importance of analyzing social, economic, and political factors, they could have emphasized the relevance of identifying how these technologies may affect not only a particular woman, but also women as a group. Assessors could have recognized that there are some harms that achieve the status of moral wrongs only when considering groups and not individuals. For instance, a particular woman may have no specific right to participate in the testing of a drug. However, if such testing excludes a group of people on account of their gender, that may be an ethical wrong because of unfair discrimination.[57] Similarly, evaluators could have stressed the significance of women's participation for IVF practice and research. Thus, although women are most affected by the new reproductive technologies, they are virtually absent from the legal, scientific, and governmental bodies that are involved in the investigation, development, and delivery of these procedures. As a result, policymakers might neglect issues that affect women.

1.7. Some Objections and Responses

My criticism of how IVF analysts have dealt with the problem of free informed consent may raise several objections. First, to my argument that

57. Dee, for example, S.M. Wolf, "Introduction: Gender and Feminism in Bioethics," in *Feminism and Bioethics*, ed., S.M. Wolf (New York: Oxford University Press, 1996), pp.3-43.

assessors have failed because they have neglected problems (underestimating insufficiency of data on IVF, overemphasizing the benefits) that could compromise women's abilities to give free informed consent, critics may object in several ways. Opponents may claim that such an argument is incorrect because, even if assessors have neglected the limitation of information on IVF safety and effectiveness and have accentuated the benefits, it does not follow that evaluators have failed to promote informed consent because they do clearly advise doctors to obtain written informed consent forms from IVF patients. Second, to my argument that assessors have erred because they have failed to address social influences that make hinder women's opportunities to give free informed consent, critics may respond in the following way. They may say that the task of IVF assessors was to analyze IVF and related procedures, not to offer value judgments about the context in which these technologies exist. I shall treat these objections in order.

1.7.1. OBTAINING WRITTEN INFORMED CONSENT FORMS

The first criticism seems particularly compelling because it stresses the importance of informed consent forms. However, the objection is incomplete because it seems to equate written consent with autonomous authorization. Thus, this criticism appears to mix two different conceptions of informed consent: informed consent as autonomous authorization and informed consent as a legal or institutional authorization.[58] Although there is a relationship between the two meanings of 'informed consent', there is a gap between them. In the first sense, "informed consent" is analyzable as an autonomous authorization. In this sense, a person must do more than declare agreement or comply with a proposal or arrangement. She or he must authorize the proposal through an act of informed and voluntary consent. An informed consent in this first sense occurs if and only if a patient or subject, with substantial understanding and in substantial absence of control by others, intentionally authorizes a professional to do something. In the second sense, informed consent is analyzable in terms of the social rules for

58. The analysis of the two senses of informed consent relies mainly on Faden and Beauchamp, *History*, part III.

institutions that must obtain legally valid consent from the patients or subjects before initiating a therapeutic procedure or research. "Informed consent" in this sense does not refer to autonomous authorization, but to a legally or institutionally effective authorization, as determined by prevailing rules.

Current literature in bioethics suggests that autonomous choices of patients and subjects must be the root for fashioning the institutional and policy requirements for effective consent.[59] Thus, whether a particular set of requirements for informed consent (in the second sense) is morally acceptable must depend in large measure on the extent to which it serves to maximize the likelihood that the conditions of informed consent in the first sense will be satisfied.

In practice, however, a carefully delineated conception of autonomous decisionmaking has generally not being the basis for institutional rules governing effective authorization. Hence, in the case of IVF treatment, practice shows that written consent may be far from guaranteeing an autonomous authorization. A study done by the Canadian Royal Commission on New Reproductive Technologies shows that there is no uniformity in IVF programs information and procedures for obtaining informed consent.[60] In some cases the materials are difficult to understand, data are lacking, and the consent forms vary widely from clinic to clinic. For example, research shows that patient information materials on IVF are often difficult to comprehend because they are written in a very technical way.[61] According to the patients, this style of writing normally conveys the message that they cannot hope to understand the complicated treatment, but should in any case comply with the directions of clinic staff. Similarly, according to Canadian Royal Commission, the majority of patients surveyed were dissatisfied with the information received when undergoing IVF treatment. Some of the

---

59. See U.S. National Commission; J. Katz, *The Silent World of Doctor and Patient* (New York: The Free Press, 1984); Faden and Beauchamp, *History*; and Beauchamp and Childress, *Principles*, ch. 3.

60. Royal Commission on New Reproductive Technologies, *Proceed with Care* ((Ottawa, Canada: Canada Communications Group, 1993), p. 547 (hereafter cited as RCNRT, *Care*).

61. RCNRT, *Care*, p. 548.

information that patients need to make informed decisions about infertility interventions includes the nature and objectives of the procedure; whether there are feasible alternatives to it; the personal chances of having a child as a result of the treatment; the nature and probability of the known and possible long-term effects of the technique; emotional demands that the treatment imposes (timing of sexual relations, frequent visits to the clinic, failing to achieve a pregnancy); the costs of the treatment; and the short-term consequences of the treatment. Very few programs address all the aspects mentioned. Most of the information materials mentioned adoption as an alternative to IVF but did not give sources of information about it. Some IVF programs, as the Canadian study assessment shows, do not inform patients about the costs they would incur when undergoing IVF treatments.[62]

Likewise, the Canadian investigation affirms that there are no standard procedures for obtaining informed consent about IVF and related procedures. Some IVF clinics have detailed written consent forms for each stage or procedure in the IVF process. Others gain consent for only some of the steps of the treatment. Also, consent forms are often difficult understand. In addition, only in some cases do doctors inform their patients that they can withdraw from the treatment at any time.[63]

Investigations done in Britain also have found that the consent forms given to women undergoing IVF treatment seem far from promoting an autonomous authorization.[64] These forms do not spell out what the treatment involves. Professionals do not tell women that they can opt out of treatment whenever they want. The written information produced by most clinics is, according to the mentioned study, a glossy brochure with pictures of medical facilities and portraits of the medical team. Also doctors ask women to decide on the future use (storage, destruction, donation, or for research) of their spare pre-embryos at the same time as they agree to undergo treatment. Some clinics give women almost no time to consider their decision. A few ask patients to sign the consent form at the initial consultation.

---

62. RCNRT, *Care*, p. 547.

63. RCNRT, *Care*, p. 548-49.

64. See N. Pfeffer, "The Uninformed Conception," in *New Scientist*, 131: 1778 (July 1991): 40-41 (hereafter cited as Pfeffer, Uninformed).

Other clinics in Britain ask for a signature just before treatment commences. In some cases, and in spite of the research indicating that hormones cause cancer, clinics request the signature at the stage of egg collection, which means that women do not give written consent to the fertility hormones that doctors use to induce ovulation.[65]

If the previously mentioned studies on the practice of obtaining informed consent for IVF treatment are correct, then signing a written consent form (as assessors require) does not guarantee the autonomy of women's decisions. Because IVF evaluations are essential steps in the selection of a public policy, if assessors neglect requirements of disclosure by underestimating uncertainty, and by overemphasizing IVF benefits, then the formulation of institutional policies for informed consent would likely be far from conforming to standards of consent in the sense of autonomous authorization. Thus, because many women may lack adequate information and understanding about IVF treatment, they may be giving a legal or institutional consent to IVF without providing free informed consent in the sense of autonomous authorization.

1.7.2. VALUE-NEUTRAL IN VITRO FERTILIZATION ASSESSMENTS

The second objection, that the assessors' role is to evaluate IVF and related technologies and not to offer value judgments about the context in which these technologies exist, is significant because it focuses on the aim of the assessments. Nevertheless, this criticism is questionable because it seems to suggest that the evaluation of a technology should disregard the social context in which people are going to implement or use it. But such a suggestion is surely wrong. IVF assessments are fundamental steps in the adoption of particular public policies concerning infertility treatments. The social context in which those public policies take place is undoubtedly essential to the evaluation of reproductive technologies. To affirm the contrary is to neglect the fact that technology assessments are not value neutral.

IVF studies are not value neutral because, as I said in chapter two, all research includes contextual and constitutive values and some also

65. See Pfeffer, Uninformed.

contain bias values.[66] Bias values are deliberate misinterpretations and omissions to serve one's own interests. Falsification of data, manipulation of statistics or interpretation of data to support one's owns prejudices are examples of bias values. Cases of scientific fraud constitute clear representations of these kinds of values.[67] Constitutive values are the source of the rules determining acceptable scientific method or practice. What counts as evidence for a theory or as an explanation of a phenomenon depends on certain assumptions or value judgments about scientific method. Some examples of constitutive values are the empirical adequacy of the theories, truth, consistency with accepted theories in other domains, objectivity, and simplicity.[68] Finally, contextual values are social, personal, and cultural preferences. They have an impact on the kind of scientific research pursued and the method employed. Contextual values also can motivate the acceptance of global assumptions that decide the character of research in an entire field.[69]

Evaluations of IVF and related procedures are not value-neutral because, although assessors can and should avoid bias values, it is in principle impossible to elude constitutive values, and it is in practice impossible to avoid contextual values.[70] First, all investigation is unalterably theory-laden in requiring both a definition of the research

---

66. See H. Longino, *Science as Social Knowledge* (Princeton, NJ: Princeton University Press, 1990) (hereafter cited as Longino, *Science*) and K. Shrader-Frechette, *Science Policy, Ethics, and Economic Methodology* (Boston: Reidel, 1985), pp.67-72 (hereafter cited as: Shrader-Frechette, *Science Policy*).

67. See, for example, S. J. Gould, *The Mismeasure of the Man* (New York: Norton, 1981); Committee on Government Operations, *Are Scientific Misconduct and Conflicts of Interest Hazardous to our Health?* (Washington, D.C.: U.S. Government Printing Office, 1990); Committee on Energy and Commerce, *Scientific Fraud* (Washington, D.C.: U.S. Government Printing Office, 1992).

68. See, for example, C. G. Hempel, *Philosophy of Natural Science* (Englewood Cliffs, NJ: Prentice Hall, 1966); K. Popper, *The Logic of Scientific Discovery*, London: Hutchinson, 1959); T. Kuhn, *The Structure of Scientific Revolutions* (Chicago: University of Chicago Press, 1962, 1970) (hereafter cited as Kuhn. *Structure*).

69. See Longino, *Science*, pp.86-89. See, also, H. Longino, "Biological Effects of Low Level Radiation: Values, Dose-Response Models, Risk Estimates," *Syntheses* 81 (1990): 391-404.

70. See H. Longino, *Science* and Shrader-Frechette, *Science Policy*, pp. 67-72.

problem and a criterion for relevant evidence.[71] Use of evaluative definitions and criteria means that the investigation can never be neutral in the sense of avoiding constitutive values. Second, contextual values often fill the gap resulting from limitation of knowledge. Because incomplete evidence limits any research, ignorance about a particular phenomenon provides an opportunity for the determination of information using contextual (social and moral) values.[72]

Existence of epistemological value judgments is not, however, the only problem for those who defend the value neutrality of IVF assessments. IVF evaluations also include central ethical assumptions, such as that infertility is a medical problem that requires a technological solution, that having genetically related children is important, and that social influences are not relevant when considering autonomy. To presuppose that IVF assessments are value neutral is then questionable because it may mask key assumptions that need to be assessed.

Presupposing that IVF assessments can and should be value neutral is also problematic because it may sanction the status quo. Hence, if assessors ignore social and political contexts of IVF implementation and use, on grounds that they have to be neutral, then their silence about existing problems or evils in our society will simply serve to legitimize whatever policy is in practice. Thus, if evaluators do not present critical remarks on the lack of data on IVF risks or on the extensive use of a technology that evidence has not proved effective, then they are implicitly sanctioning the current state of affairs in the practice of IVF. Thus, many women may continue using seriously risky and ineffective infertility treatments without knowing it. As a consequence, assessors may encourage policies where women may be put at risk without their consent. Similarly, in failing to analyze critically the use of IVF, evaluators may promote discriminatory practices against women where they undergo ineffective treatments because their partners are infertile. Also, in disregarding social, political, and economic factors, assessors may be condoning practices that obstruct voluntary decisionmaking.

---

71. See, for example, N. R. Hanson, *Patterns of Discovery* (Cambridge: Cambridge University Press, 1958); Michael Polanyi, *Personal Knowledge* (New York: Harper and Row, 1958); Kuhn, *Structure*.

72. See H. Longino, *Science* and Shrader-Frechette, *Science Policy*, pp. 71-72.

Furthermore, ignoring the fact that the evaluation of IVF is not value neutral may lead to overemphasizing the technical aspects of public policy concerning infertility and to neglecting ethical and political solutions to reproductive difficulties. Such a consequence is problematic because ethical and political solutions may help more people. We should not forget that IVF assessments are fundamental steps in the selection of public policies. And public policy affects all of us. If assessors explicitly analyze their ethical and epistemological assumptions, they would more likely help policymakers in their task because they would take into account the impacts of different policy options. Assessors would also help citizens to exercise democratic control over policies that go against their values.

1.8. Summary and Conclusion

Because one of the main goals of the four IVF analyses is to advise governments in the selection of sound public policies on infertility treatments, they should have given special consideration to the problem of free informed consent. However, I have argued that assessors of IVF have failed to do so. As a result, their evaluations may encourage policies that jeopardize women's rights to free informed consent. First, I have claimed that because assessors have failed to address adequately problems with disclosure of information, their analyses may compromise women's abilities to give free informed consent. Evaluators have overlooked questions of disclosure of information because (i) they have underestimated the lack of scientific evidence on IVF safety, (ii) they have undervalued difficulties with the presentation of IVF success rates, and (iii) they have overemphasized the benefits of the procedures and have downplayed the hazards. Hence, analysts have skewed the balance of pros and cons and have made it difficult for women to give informed consent to a risky treatment whose effects are uncertain. Second, I have argued that assessors have neglected problems with voluntariness. Because evaluators have failed to analyze appropriately social and economic conditions in which women make decisions about IVF, they have overlooked circumstances that could defeat women's rights to free informed consent.

I do not wish to imply with my analysis of free informed consent that social, political, and economic circumstances always render women's decisions about infertility treatments nonvoluntary. My point has been to criticize IVF assessments for neglecting the evaluation of factors that could restrict women's opportunities to give free informed consent.

Free informed consent requirements are, without doubt, a landmark in the history of rights of patients and research subjects. But if so, the evaluation of medical technologies should pay special attention to those explicit factors, such as adequate disclosure of risks and benefits, that may compromise people's rights to give free informed consent. If this chapter is correct, assessors also should take into account the implicit social, political, and economic factors that may hinder people's opportunities to make voluntary decisions.

CHAPTER 7

CONCLUSION AND FUTURE TRENDS

1.1. Introduction

New assisted-conception technologies such as IVF are changing the context of human reproduction. Many people now experience new hopes of overcoming infertility and having children. However, these new possibilities also create important ethical and policy problems and have the potential to cause conflicts among different parties and interests. Some of the issues related to reproductive procedures have reached the public policy agenda (e.g., licensing of clinics, regulations about embryo research and frozen embryos), but many other important matters (e.g., informed consent issues, social impacts of IVF and related techniques) are still unexplored.

This work has tried to show how inadequate technology assessments may have caused questionable public policies in relation to IVF and associated procedures. I have argued that in evaluating these technologies, assessors have treated epistemological and ethical issues--such as the significance of choosing criteria for deciding in situations of uncertainty, the importance of individuals' rights over the common good, and the relevance of the social context for free informed consent--in ways that may be problematic. For example, in spite of the scientific uncertainty surrounding IVF effectiveness and risks, assessors have preferred to minimize false positives. As a consequence, they have sanctioned the largely unrestricted use and expansion of IVF. Similarly, in framing the infertility question as mainly an individual problem, evaluators have neglected the impacts of these technologies on the common good. Likewise, assessors have ignored social, political, and economic issues that could jeopardize women's rights to free informed consent.

1.2. Some Consequences of This Analysis

History reveals the dangers of implementing technologies that are inadequately assessed. Pesticides, thalidomide, nuclear power, and DES offer some examples of the dangers involved. Granted, evaluation of new techniques is often difficult because it requires assessors to determine current desirable actions in accordance with possible future effects. However, some of the deficiencies of technology assessments (i.e., the ones I have analyzed in this work) may be avoidable with a more complete approach to the analysis of new and existing techniques. Thus, if my examination of IVF assessments is correct, some consequences follow for philosophy and public policy that should influence future research in both areas.

1.2.1. IMPACTS ON PHILOSOPHY

One of the consequences of this work for philosophy is that, if my analysis is correct, then philosophers have a responsibility to offer new approaches to understanding technology. As I said in chapter one, philosophers are specially trained for conceptual clarification, valid argumentation, exposing hidden assumptions, comprehending moral theories, and critically examining consequences. This training gives them an opportunity to contribute to a better understanding of technology assessment. Although these skills are not sufficient to guarantee an adequate evaluation or decision, they can help in providing a more comprehensive one. Thus, philosophers may contribute to evaluating medical and other procedures by recasting the way we define problems, by exploring the relationships between technical and nontechnical matters, and by analyzing technology itself as problematic. For example, ethical problems related to experimentation on human embryos have given place to the emergence of the concept of "pre-embryo" in order to avoid ethical difficulties in research. This term seeks to establish a morally relevant difference between an implanted embryo, that has the potential to become a child, and a pre-implanted embryo that lacks that potential.

Philosophers may also contribute to the assessment of technologies by analyzing ways in which technological developments may be responsible for the creation of problems that did not exist before the implementation of new procedures. For instance, prior to the massive introduction of the automobile, city pollution was not a question of concern. Similarly, a few decades ago, having a child with Down's Syndrome appeared to be something that parents could not prevent. Now, however, the availability of prenatal testing and selective abortion may contribute to decisionmaking about potential Down's Syndrome children. Examination of how technological innovations create or transform particular problems is then relevant for adequate assessments because it may be able to clarify moral debates.

Philosophers also can contribute to better assessments by studying the interconnections between technology and society. Technology exists within networks of social practices. For example, introduction of hormonal contraceptives would be ineffective in a culture where women have limited access to health-care institutions. An evaluation that takes into account epistemological and ethical problems of those networks is thus essential for understanding the moral dimensions of new procedures.[1]

An analysis of social, political, and economic factors that may hinder free informed consent could be another contribution of philosophers to a better assessment of technologies. Philosophers could evaluate those circumstances such as situations of dependency, discrimination, and economic duress that may obstruct people's free and informed decisionmaking.[2] Such inquiry would likely work to empower those in disadvantaged positions in our society.

Philosophers should also promote analysis of new and existing technologies that take into account gender issues. Too often ethical and political theorists have presented human beings as genderless (or simply as males). This image may have blinded philosophers to specific problems concerning women. For instance, women are more often patients than males

---

1. See, for example, MA. Warren, *Gendercide: The Implications of Sex-Selection* (Totowa, NJ: Rowman and Allanheld, 1985); L. Winner, *The Whale and the Reactor* (Chicago: The University of Chicago press, 1986); S. Jasanoff, *The Fifth Branch: Science Advisors as Policymakers* (Cambridge: Cambridge University press, 1990); K.H. Rothenberg and E. J. Thomson (eds.), *Women and Prenatal Testing: Facing the Challenges of Genetic Technology* (Columbus, OH: Ohio State University Press, 1994).

2. See, for example, J. Feinberg, *Social Philosophy* (Englewood Cliffs, NJ: Prentice Hall, 1973); R. Faden and T. Beauchamp, *A History and Theory of Informed Consent* (New York: Oxford University Press, 1986); K. Shrader-Frechette, *Risk and Rationality* (Berkeley: University of California Press, 1991); K. Shrader-Frechette, Burying Uncertainty (Berkeley: University of California Press, 1993); P. Singer, *Practical Ethics* (Cambridge: Cambridge University Press, 1993).

are, and they have more physician contact. However, physicians providing care are mostly men. Hence, in emphasizing gender issues, philosophers may help to contextualize medical and scientific practices that have harmed women. For instance, some of the more important controversies in bioethics have been about reproductive issues. Ethical and legal rules about all these issues will affect women most. Hence, an analysis of such rules should take into account gender inequalities they may reinforce. Similarly, stressing gender issues may reveal the significance of cases such as DES, breast implants, IUDs, unnecessary and forced cesareans, needless hysterectomies, and the exclusion of women and women's health problems from research. A theoretical approach that directs attention to gender problems would likely offer new perspectives on discriminatory practices against women and foster a revision of the roles of medicine and science in our society.[3]

1.2.2. REPERCUSSIONS FOR PUBLIC POLICY

If my work on the inadequacies of IVF assessments is correct, several consequences also would follow for public policy. The most obvious is the need for a reassessment of IVF and related technologies. The extensive use of these procedures may have effects on our society too problematic to leave untreated.

As a way to solve some of the problems (i.e., lack of information) in IVF assessments, policymakers should give priority to funding research on the demography, causes, and treatment options of infertility. Equally necessary is the systematic evaluation of the short- and long-term risks and effectiveness of IVF. Without adequate data on hazards and benefits, a sound public policy on infertility treatments such as IVF seems unlikely. Until knowledge in these areas is adequate, policymakers should take measures to prevent unethical experimentation on women. For example, they could promote services (independent of infertility clinics) that offer counseling and information about risks, benefits, and costs of different available options to people facing reproductive problems.

---

3. See, for example, S.M. Wolf, ed., *Feminism and Bioethics* (New York: Oxford University Press, 1996).

Another consequence for public policy that follows from my analysis of IVF assessments is the need for considering measures that could treat, prevent, and dissolve the problem of infertility. A broader approach to reproductive difficulties would likely help people to confront their problems by giving infertile people a variety of options, from preventive measures to technological treatments. Hence, they would have more opportunities for making meaningful choices.

Also, following from my evaluation of the inadequacies of IVF studies, policymakers should consider the creation of public mechanisms to protect the interests of all parties involved in IVF, especially those most vulnerable to exploitation and abuse, such as poor and minority women. Policymakers also should be concerned with the proposal of strategies that favor women's and minorities' participation in public policy, biotechnology research, and medical practice. Such participation would likely increase the chances of these groups to exercise their rights to free and informed consent.

1.3. Conclusion

If we do not properly assess and manage technology, it can endanger our democratic institutions, harm particular individuals, and separate societal power and responsibility. Too many important things, such as rights to free informed consent, social stability, rights to bodily security, and the well-being of our communities, are at stake to disregard the importance of adequately assessing new and existing technologies.

INDEX